W9-CMF-711

ALSO BY MATTHEW B. CRAWFORD

Shop Class as Soulcraft: An Inquiry into the Value of Work

THE WORLD BEYOND YOUR HEAD

THE WORLD BEYOND YOUR HEAD

ON BECOMING AN INDIVIDUAL IN AN AGE OF DISTRACTION

Matthew B. Crawford

ALLEN LANE

ALLEN LANE
an imprint of Penguin Canada Books Inc., a Penguin Random House Company

Published by the Penguin Group
Penguin Canada Books Inc., 90 Eglinton Avenue East, Suite 700, Toronto, Ontario, Canada M4P 2Y3

Penguin Group (USA) LLC, 375 Hudson Street, New York, New York 10014, U.S.A.
Penguin Books Ltd, 80 Strand, London WC2R 0RL, England
Penguin Ireland, 25 St Stephen's Green, Dublin 2, Ireland (a division of Penguin Books Ltd)
Penguin Group (Australia), 707 Collins Street, Melbourne, Victoria 3008, Australia
(a division of Pearson Australia Group Pty Ltd)
Penguin Books India Pvt Ltd, 11 Community Centre, Panchsheel Park, New Delhi – 110 017, India
Penguin Group (NZ), 67 Apollo Drive, Rosedale, Auckland 0632, New Zealand
(a division of Pearson New Zealand Ltd)
Penguin Books (South Africa) (Pty) Ltd, 24 Sturdee Avenue, Rosebank, Johannesburg 2196, South Africa

Penguin Books Ltd, Registered Offices: 80 Strand, London WC2R 0RL, England

First published in Allen Lane hardcover by Penguin Canada Books Inc., 2015. Simultaneously published in the U.S.A. by Farrar, Straus and Giroux, 18 West 18th Street, New York, NY 10011.

1 2 3 4 5 6 7 8 9 10 (RRD)

Grateful acknowledgment is made for permission to reprint excerpts from the following previously published material:

To Princeton University Press for permission to cite excerpts from Natasha Schüll's *Addiction by Design: Machine Gambling in Las Vegas* (2012).
To The Onion for permission to reprint "Man on Cusp of Having Fun Suddenly Remembers Every Single One of His Responsibilities." Reprinted with permission of The Onion. Copyright © 2014 by Onion, Inc. www.theonion.com.

Designed by Jonathan D. Lippincott
Manufactured in the U.S.A.

LIBRARY AND ARCHIVES CANADA CATALOGUING IN PUBLICATION

Crawford, Matthew B., author
The world beyond your head : on becoming an individual
in an age of distraction / Matthew B. Crawford.

Includes bibliographical references and index.
ISBN 978-0-670-06639-1 (bound)

1. Self--Social aspects. 2. Individuality--Social aspects.
3. Distraction (Psychology). 4. Mental health. 5. Civilization,
Modern--21st century--Psychological aspects. I. Title.

BF697.5.S65C73 2015 155.2 C2015-900333-4

---

eBook ISBN 978-0-14-319446-0

American Library of Congress Cataloging in Publication data available

Visit the Penguin Canada website at **www.penguin.ca**
Special and corporate bulk purchase rates available; please see
www.penguin.ca/corporatesales or call 1-800-810-3104

The great thing is to gather new vigor in reality.

—Vincent van Gogh

CONTENTS

PREFACE

We are living through a crisis of attention that is now widely remarked upon, usually in the context of some complaint or other about technology. As our mental lives become more fragmented, what is at stake often seems to be nothing less than the question of whether one can maintain a coherent self. I mean a self that is able to act according to settled purposes and ongoing projects, rather than flitting about. Because attention is so fundamental to our mental lives, this widely felt problem presents a rare occasion when an entire society is compelled to ask anew a very old question: What does it mean to be human?

Such a reconsideration has been made necessary by profound cultural changes. I find that these changes have a certain coherence to them, an arc—one that begins in the Enlightenment, accelerates in the twentieth century, and is perhaps culminating now. Though digital technologies certainly contribute to it, our current crisis of attention is the coming to fruition of a picture of the human being that was offered some centuries ago. This picture is so pervasive that it is difficult to make an object of scrutiny. At the center of it is a certain understanding of how a person encounters the world beyond his or her head.

We are said to do so only through our mental *representations* of the world. Life then imitates theory: Ours is now a highly mediated existence in which, sure enough, we increasingly encounter the world through representations. These are manufactured for us. Human experience has become a highly engineered and therefore manipulable thing.

My efforts to understand the experience of attending to real objects and to other people have led me to call our founding doctrines of human cognition into question, and to investigate the pressures they put on everyday life. They do so by rendering some aspects of our own experience illegible to us. In the course of this inquiry, some of the strangeness of our culture—for example, our approach to education and the mood of our public spaces—comes into focus.

Drawing on certain dissident strands of thought in the philosophical tradition, I offer what I take to be a more adequate picture of how we encounter objects and other people. My hope is that this alternative understanding can help us think clearly about our current crisis of attention, and reclaim certain possibilities of human flourishing.

The weight of this positive argument is carried by case studies of attention in various skilled practices. The point of these is not to spur the reader to fantasize about becoming a short-order cook, a motorcycle racer, or a builder of pipe organs. Rather, activities like these, which elicit complete immersion in a particular situation, reveal something about our constitution that tends to get lost in the official self-understanding of the West.

Skilled practices serve as an anchor to the world beyond one's head—a point of triangulation with objects and other people who have a reality of their own. The most surprising thing to emerge in this inquiry (for me, at least) is that through such triangulation we may achieve something like "individuality." For it *is* an achievement, especially in a mass society that speaks an idiom of individualism and thereby obscures the genuine article.

THE WORLD BEYOND YOUR HEAD

INTRODUCTION: ATTENTION AS A CULTURAL PROBLEM

The idea of writing this book gained strength one day when I swiped my bank card to pay for groceries. I watched the screen intently, waiting for it to prompt me to do the next step. During the following seconds it became clear that some genius had realized that a person in this situation is a captive audience. During those intervals between swiping my card, confirming the amount, and entering my PIN, I was shown advertisements. The intervals themselves, which I had previously assumed were a mere artifact of the communication technology, now seemed to be something more deliberately calibrated. These haltings now served somebody's interest.

Such intrusions are everywhere. Taking a flight recently to Chicago, I pulled down the tray from the seat back in front of me and discovered that the entire tray top was devoted to an advertisement for Droid, the multimedia smartphone. At O'Hare International Airport, the moving handrail on the escalator was covered with an endlessly recurring message from the Lincoln Financial Group: You're In Charge.® When I got to my hotel, I was handed a key card that was printed on one side with an advertisement for Benihana, the restaurant. Somehow, the fact that such a key card presents about five square inches for inevitable eyeballing had gone unnoticed, or rather unmonetized, until recently. Capitalism has gotten hip to the fact that for all our talk of an information economy, what we really have is an attentional economy, if the term "economy" applies to what is scarce and therefore valuable. As these last examples illustrate, the pertinent development here is a social technology, not something electronic. Turning unavoidable public surfaces into sites of marketing isn't inherently "digital."

We have developed methods for tuning out commercial messages, for example by inserting earbuds or burying our faces in our devices. Bus riders in Seoul, South Korea, find themselves at a new frontier: they have advertising squirted into their noses. A smell resembling that of Dunkin' Donuts coffee is released into the ventilation system as a Dunkin' Donuts advertisement plays over the bus's sound system shortly before the bus stops outside a Dunkin' Donuts store. An announcer points out the fact, in case it has somehow been missed. This kind of advertising is especially aggressive and indiscriminate, yet is also exquisitely well targeted to morning commuters who are primed to want coffee at the time they are exposed to the advertising, and there it is, right next to the bus stop! The advertising agency responsible was rewarded by its peers with a Bronze Lion award for "best use of ambient media."[1]

There remain many areas for further progress. The homework, report cards, permissions slips, and other minor communications that a teacher sends home with students are in many school districts still blank on the back. Here is a gross offense against the efficient use of space. One forward-thinking school district in Peabody, Massachusetts, now sells advertising space on the backs of these slips of paper.

But intrusive advertising is just the tip of a larger cultural iceberg; some of the positive attractions of our attentional environment are no less troubling than the unwanted aspects. It's hard to open a newspaper or magazine these days without reading a complaint about our fractured mental lives, diminished attention spans, and a widespread sense of distraction. Often the occasion for such a story is some new neuroscience finding about how our brains are being rewired by our habits of information grazing and electronic stimulation. Though it is in the first place a faculty of individual minds, it is clear that attention has also become an acute collective problem of modern life—a cultural problem.

Our susceptibility to being buffeted by various claims on our attention is surely tied to the "intensification of nervous stimulation" that the German sociologist Georg Simmel identified with the metropolitan environment over a hundred years ago. Think of the corporate manager who gets two hundred emails per day and spends his time responding pell-mell to an incoherent press of demands. The way we experience this, often, is as a crisis of self-ownership: our attention isn't simply ours to direct where we will, and we complain about it bitterly. Yet this same person may find himself checking his email frequently once he gets home or while on vacation. It becomes effortful for him to be fully present while giving his children a bath or taking a meal with his spouse. Our changing technological environment generates a need for ever more stimulation. The *content* of the stimulation almost becomes irrelevant. Our distractibility seems to indicate that we are agnostic on the question of what is *worth* paying attention to—that is, what to value.[2]

To answer this question freely requires shelter; a space for seriousness. The moralist will say that one has to carve out this space for oneself resolutely, against the noise, and that to fail to rise to this task of evaluation is to give oneself over to nihilism, in which all distinctions are leveled and all meaning gives way to mere "information."

A sociologist might go easier on us and locate our difficulty not in our individual moral failures but in a collective situation, pointing out that there aren't many limits on our mental lives of the sort that prevailed before we had immediate access to the world beyond our own narrow horizon of experience. That horizon has been exploded; all manner of once-weird stuff is now a click away. There are so many enticements, but just as important, there is little in the way of authoritative guidance of the sort that was once supplied by tradition, religion, or the kind of communities that make deep demands on us.

The moralist and the sociologist are both right. The question of what to attend to *is* a question of what to value, and this question is no longer answered for us by settled forms of social life. We have liberated ourselves from all that. The downside is that as autonomous individuals, we often find ourselves isolated in a fog of choices. Our mental lives become shapeless, and more susceptible to whatever presents itself out of the ether. But of course these presentations are highly orchestrated; commercial forces step into the void of cultural authority and assume a growing role in shaping our evaluative outlook on the world. Because of the scale on which these forces operate, our mental lives converge in a great massification—ironically, under the banner of individual choice.

Our mental fragmentation can't *simply* be attributed to advertising, the Internet, or any other identifiable villain, for it has become something more comprehensive than that, something like a style of existence. It is captured pretty well in the following satirical news item from *The Onion*.

> GAITHERSBURG, MD—While cracking open his second beer as he chatted with friends over a relaxed outdoor meal, local man Marshall Platt, 34, was reportedly seconds away from letting go and enjoying himself when he was suddenly crushed by the full weight of work emails that still needed to be dealt with, . . . an upcoming wedding he had yet to buy airfare for because of an unresolved issue with his Southwest Rapid Rewards account, and phone calls that needed to be returned.
>
> "It's great to see you guys," said the man who had been teetering on the brink of actually having fun and was now mentally preparing for a presentation that he had to give on Friday and compiling a list of bills that needed to be paid before the 7th. "This is awesome."
>
> "Anyone want another beer?" continued Platt as he reminded himself to pick up his Zetonna prescription. "Think I'm gonna grab one."
>
> Platt, who reportedly sunk into a distracted haze after coming to the razor's edge of experiencing genuine joy, fully intended to go through the motions of talking with friends and appearing to have a good time, all while he

> mentally shopped for a birthday present for his mother, wracked his brain to remember if he had turned in the itemized reimbursement form from his New York trip to HR on time, and made a silent note to call his bank about a mysterious recurring $19 monthly fee that he had recently discovered on his credit card statement.[3]

I think most of us can recognize ourselves in Mr. Platt. Is "modern life" really so burdensome? Yes it is. But Mr. Platt seems to have a deeper difficulty as well: joy can get no grip on him. The sketch seems to be about the little tasks that claim his attention, but at the center of it is an ethical void. He is unable to actively affirm as important the pleasure of being with friends. He therefore has no basis on which to resist the colonization of life by *hassle.*

Clearly, no single discipline or body of thought is adequate to parse the crisis of attention that characterizes our cultural moment. There is a rich literature on attention in cognitive psychology, extending from William James's work of a century ago to the latest findings in childhood development. There are scattered treatments in moral philosophy, and these are indispensable. The fact has not been widely noticed, but attention is the organizing concern of the tradition of thought called phenomenology, and this tradition offers a bridge between the mutually uncomprehending fields of cognitive psychology and moral philosophy. What is required, then, is a highly synthetic effort—we can call it philosophical anthropology.

Through this inquiry I hope to arrive at something like an *ethics of attention* for our time, grounded in a realistic account of the mind and a critical gaze at modern culture. I should note here that I am using the term "ethics" in its original sense—not primarily as an account of what we are obliged or forbidden to do, but as a more capacious reflection on the sort of *ethos* we want to inhabit. Nor do I wish to join the culture wars surrounding "technology"—as being either an apocalyptic force or a saving one that heralds the arrival of a new global intelligence, etc. I want rather to tunnel beneath that intellectual cul-de-sac and trace the subterranean strata—the historically sedimented geological structures—of our age of distraction, the better to map our way out of it.

An ethics of attention would have to begin by taking seriously, and trying to make sense of, the qualitative character of first-person experience in our contemporary cognitive environment: by turns anxious, put-upon, distracted, exhausted, enthralled, ecstatic, self-forgetting. The thing is, we are very sophisticated. As the inheritors of layers of theorizing about the human person, we find it no trivial task to recover a more direct access to our own experience. In the course of trying to do that, I have found it necessary to scrutinize certain background assumptions about the self that shape our experience. It has been said (by Iris Murdoch) that man is the animal that makes pictures of himself, and then comes to resemble the pictures.

Such pictures come to us from various departments of the human sciences. In ways that bear directly on our theme, these sciences continue to be informed by the agenda of the Enlightenment. (I will have more to say about that shortly.) This agenda shaped a very partial view of the human person, one that we have been operating with for centuries but has become in various ways poorly suited to our circumstances. My hope is that a fuller picture will be both truer and more serviceable for us in finding a way through our current predicament of attention.

But I have gotten ahead of the argument. Allow me to simply describe some further dimensions of that predicament.

THE ATTENTIONAL COMMONS

We have all had the experience of sitting in an airport with an hour to kill and being unable to escape the chattering of CNN. The audio may be turned off, but if the TV is within view, I, for one, find it impossible not to look at it. The introduction of novelty into one's field of view commands what the cognitive psychologists call an orienting response (an important evolutionary adaptation in a world of predators): an animal turns its face and eyes toward the new thing. A new thing typically appears every second on television. The images on the screen jump out of the flow of experience and make a demand on us. In their presence it is difficult to rehearse a remembered conversation, for example. Whatever trains of thought might otherwise be pursued by those in the room give way to a highly coordinated experience: not the near-simultaneous turning of a troupe of macaques to face the python that has appeared, but the involuntary glances of weary travelers toward the "content" on offer.

Alternatively, people in such places stare at their phones or open a novel, sometimes precisely in order to tune out the piped-in chatter. A multiverse of private experiences is accessible after all. In this battle of attentional technologies, what is lost is the kind of *public* space that is required for a certain kind of sociability. Jonathan Franzen wrote, "Walking up Third Avenue on a Saturday night, I feel bereft. All around me, attractive young people are hunched over their StarTacs and Nokias with preoccupied expressions, as if probing a sore tooth . . . All I really want from a sidewalk is that people see me and let themselves be seen . . ."

A public space where people are not self-enclosed, in the heightened way that happens when our minds are elsewhere than our bodies, may feel rich with possibility for spontaneous encounters. Even if we do not converse with others, our mutual reticence is experienced *as* reticence *if* our attention is not otherwise bound up, but is rather free to alight upon one another and linger or not, because we ourselves are free to pay out our attention in deliberate measures. To be the object of someone's reticence is quite different from not being seen by them; we may have a vivid experience of having encountered another person, even if in silence. Such encounters are always ambiguous, and their need for interpretation gives rise to a train of imaginings, often erotic. This is what makes cities exciting.

Psychologists have suggested that attention may be categorized by whether it is goal-driven or stimulus-driven, corresponding to whether it is in the service of one's own will or not. A teacher taking a head count on a chaotic school bus is engaged in the first, "executive" kind of attention. By contrast, if there is a sudden bang outside my window, my attention is stimulus-driven. I may or may not go to the window to investigate, but the claim on my attention is involuntary.

The orienting response requires of us a concerted effort of executive attention if we are to resist it, and our capacity for such resistance is finite. Of course, in my airport example, one can simply shift in one's seat and avert one's gaze from the screens. But the fields of view that haven't been claimed for commerce seem to be getting fewer and narrower. The ever more complete penetration of public spaces by attention-getting technologies exploits the orienting response in a way that preempts sociability, directing us away from one another and toward a manufactured reality, the content of which is determined from afar by private parties that have a material interest in doing so. There is no conspiracy here, it's just the way things go.

When we go through airport security, the public authority makes a claim on our attention for the common good. This moment is emblematic of *the* purpose for which political authority in a liberal regime is originally instituted—public safety—and rightly has a certain gravity to it. But in the last few years, I have found I have to be careful at the far end of the process, because the bottoms of the gray trays that you place your items in for X-ray screening are now papered with advertisements, and their visual clutter makes it very easy to miss a pinky-sized flash memory stick against a picture of fanned-out L'Oréal lipstick colors.

I am already in a state of low-level panic about departure times, possible gate changes, and any number of other contingencies that have to be actively monitored while traveling, to say nothing of the fact that my memory is tapped out with detailed concerns about the talk I am going to have to give in front of strangers in a few hours. This fresh demand for vigilance, lest I lose my PowerPoint slide show, feels like a straightforward conflict between me and L'Oréal.

Somehow L'Oréal has the Transportation Security Administration on its side. *Who* made the decision to pimp out the security trays with these advertisements? The answer, of course, is that Nobody decided on behalf of the public. Someone made a suggestion, and Nobody responded in the only way that seemed reasonable: here is an "inefficient" use of space that could instead be used to "inform" the public of "opportunities." Justifications of this flavor are so much a part of the taken-for-granted field of public discourse that they may override our immediate experience and render it unintelligible to us. Our annoyance dissipates into vague impotence because we have no public language in which to articulate it, and we search instead for a diagnosis of our*selves*: Why am I so angry? It may be time to adjust the meds.

In the main currents of psychological research, attention is treated as a resource—a person has only so much of it. Yet it does not occur to us to make a claim for our attentional resources on our own behalf. Nor do we yet have a political economy corresponding to this resource, one that would take into account the peculiar violations of the modern cognitive environment. Toward this end, I would like to offer the concept of an *attentional commons*.

There are some resources that we hold in common, such as the air we breathe and the water we drink. We take them for granted, but their widespread availability makes everything else we do possible. I think the absence of noise is a resource of just this sort. More precisely, the valuable thing that we take for granted is the condition of *not being addressed*. Just as clean air makes respiration possible, silence, in this broader sense, is what makes it possible to *think*. We give it up willingly when we are in the company of other people with whom we have some relationship, and when we open ourselves to serendipitous encounters with strangers. To be addressed by mechanized means is an entirely different matter.

The benefits of silence are off the books. They are not measured directly by any econometric instrument such as gross domestic product, yet the availability of silence surely contributes to creativity and innovation. They do not show up explicitly in social statistics such as level of educational achievement, yet one consumes a great deal of silence in the course of becoming educated.

If clean air and water were no longer the rule for us, the economic toll would be truly massive. This is easy to grasp, and that is why we have regulations in place to protect these common resources. We recognize their importance and their fragility. We also recognize that absent robust regulations, air and water will be used by some in ways that make them unusable for others—not because they are malicious or careless, but because they can make money using them this way. When this occurs, it is best understood as a *transfer of wealth* from "the commons" to private parties.

A notable feature of the gangsterish regimes that rule in many formerly Communist countries is the apparent absence, or impotence, of any notion of a common good. Wherever communism was established by coercion, when it later collapsed and private interests were allowed to assert themselves it became clear that there was no well-established intellectual foundation for defending such shared resources as clean air and water. Many citizens of these countries now live in the environmental degradation that results when privatization has no countervailing force of public-spiritedness. We in the liberal societies of the West find ourselves headed toward a similar condition with regard to the resource of attention, because we do not yet understand it to be a resource.[4]

Or do we? Silence is now offered as a luxury good. In the business-class lounge at Charles de Gaulle airport, what you hear is the occasional tinkling of a spoon against china. There are no advertisements on the walls, and no TVs. This silence, more than any other feature of the space, is what makes it feel genuinely luxurious. When you step inside and the automatic airtight doors whoosh shut behind you, the difference is nearly tactile, like slipping out of haircloth into satin. Your brow unfurrows itself, your neck muscles relax; after twenty minutes you no longer feel exhausted. The *hassle* lifts.

Outside the lounge is the usual airport cacophony. Because we have allowed our attention to be monetized, if you want yours back you're going to have to pay for it.

As the commons gets appropriated, one solution, for those who have the means, is to leave the commons for private clubs such as the business-class lounge. Consider that it is those in the business lounge who make the decisions that determine the character of the peon lounge and we may start to see these things in a political light. To engage in playful, inventive thinking, and possibly create wealth for oneself during those idle hours spent at an airport, requires silence. But other people's minds, over in the peon lounge (or at the bus stop) can be treated as a resource—a standing reserve of purchasing power to be steered according to innovative marketing ideas hatched by the "creatives" in the business lounge. When some people treat the minds of other people as a resource, this is not "creating wealth," it is a transfer.[5] The much-discussed decline of the middle class in recent decades, and the ever greater concentration of wealth in a shrinking elite, may have something to do with the ever more aggressive appropriations of the attentional commons that we have allowed to take place.

This becomes especially pertinent in the era of big data, when we find ourselves the objects of attention-getting techniques that are not only pervasive, but increasingly well targeted. There is currently much talk of a right to privacy in our digital lives. Apart from the usual concerns about online security and identity theft, I have to confess that I am not terribly worried about keeping particular facts about myself hidden from the data-mongers—*until* they use that data to make a claim on my attention. I think we need to sharpen the conceptually murky right to privacy by supplementing it with a *right not to be addressed*. This would apply not, of course, to those who address me face-to-face as individuals, but to those who never show their face, and treat my mind as a resource to be harvested by mechanized means.

Attention is the thing that is most one's own: in the normal course of things, we choose what to pay attention to, and in a very real sense this determines what is real for us; what is actually present to our consciousness. Appropriations of our attention are then an especially intimate matter.

But it is also true that our attention is directed to a world that is shared; one's attention is not simply one's own, for the simple reason that its objects are often present to others as well. And indeed there is a moral imperative to *pay attention* to the shared world, and not get locked up in your own head. Iris Murdoch writes that to be good, a person "must know certain things about his surroundings, most obviously the existence of other people and their claims."[6]

Consider the person talking on his cell phone while cruising through a crowded suburban commercial district, with a motorcyclist in the lane next to him. Driving while talking on a cell phone impairs performance as much as driving while legally drunk.[7] It doesn't matter whether the phone is hands-free or not; the issue is that having a conversation uses attentional resources, of which we have a finite amount. It especially impairs our ability to notice and register novel things in the environment; psychologists call this inattentional blindness. Pedestrians who walk while talking on a cell phone weave more, change direction more, cross the street in a riskier way, are less likely to acknowledge others (that is, be sociable), and, in the findings of a recent experiment, are less likely to notice the clown on a unicycle who just rode past.[8] Put a person with this level of impairment behind the wheel of a two-ton, two-hundred-horsepower car and his blindness becomes an apt topic in discussions of what we owe one another. In the attentional commons, circumspection—literally looking around—would be one element of justice.

One of the more interesting findings to come out of the research on distracted driving is that, while having a cell phone conversation impairs driving ability, having a conversation with someone present in the car does not. A person who is present can cooperate by modulating the conversation in response to the demands of the driving situation.[9] For example, if the weather is bad he tends to be quiet. A passenger acts as another pair of eyes on the situation he inhabits with the driver, and tends to improve a driver's ability to notice and quickly respond to out-of-the-ordinary challenges.

The idea of a commons is suitable in discussing attention because, first, the penetration of our consciousness by interested parties proceeds very often by the *appropriation* of attention in public spaces, and second, because we rightly *owe* to one another a certain level of attentiveness and ethical care. The words italicized in the previous sentence rightly put us in a political economy frame of mind, if by "political economy" we can denote a concern for justice in the public exchange of some private resource.

THE ASCETICS OF ATTENTION

The existentialist writer Simone Weil and the psychologist William James both suggested that the struggle to pay attention trains the faculty of attention; it is a habit built up through practice. Grappling with a problem for which one has little aptitude or inclination (a geometry problem, say) exercises one's power to attend. For Weil, this ascetic aspect of attention—the fact that it is a "negative effort" against mental sloth—is especially significant. "Something in our soul has a far more violent repugnance for true attention than the flesh has for bodily fatigue. This something is much more closely connected with evil than is the flesh. That is why every time that we really concentrate our attention, we destroy the evil in ourselves." Students must therefore work "without any reference to their natural abilities and tastes; applying themselves equally to all their tasks, with the idea that each one will help to form in them the habit of attention which is the substance of prayer."

It should be duly noted that Weil was a mystic who (some say) deliberately starved herself to death, and indeed her dismissal of natural inclinations in the young suggests she was more infatuated with self-mortification than she was seriously concerned with how students might best learn. Yet Weil's existential melodrama shouldn't prevent us from appreciating her point that the ascetic disposition has an important role in education. To attend to anything in a sustained way requires actively *excluding* all the other things that grab at our attention. It requires, if not ruthlessness toward oneself, a capacity for self-regulation.

And reciprocally, the ability to control oneself in the face of some temptation is greatly enhanced by, indeed seems simply to *be*, the ability to direct one's attention toward something else. In a classic psychology experiment, Walter Mischel and E. B. Ebbesen gave children the option of having one marshmallow immediately or, if they were able to wait fifteen minutes, two marshmallows.[10] Left alone with the marshmallow at hand, some broke down and gobbled it immediately, others after a brief struggle. But about a third of the children succeeded in deferring gratification and getting the bigger payoff. Those who did so were those who distracted themselves from the marshmallow by playing games under the table, singing songs, or imagining the marshmallow as a cloud, for example. In a follow-up study of the same children a dozen years later, their initial performance on the self-regulation task was more predictive of life success than any other measure, including IQ and socioeconomic status. The researchers' interpretation of their results is that it isn't willpower (as conventionally understood) that distinguishes the successful children, it is the ability to strategically allocate their attention so that their actions aren't determined by the wrong thoughts. Self-regulation, like attention, is a resource of which we have a finite amount. Further, the two resources are intimately related. Thus, if someone is tasked with controlling her impulses for some extended period of time, her performance shortly thereafter on a task requiring attention is degraded.

Without the ability to direct our attention where we will, we become more receptive to those who would direct our attention where *they* will—to the omnipresent purveyors of marshmallows. To the extent that the power of concentration is widely attenuated, so too is the power of self-regulation. We become more easily suggestible and buy more stuff. I suppose this is good for economic growth. But if consumer capitalism can go on only by continuing to accelerate the "intensification of nervous stimulation," there would seem to be a fundamental antagonism between this form of economic life and the individual who inhabits it. That is, we may have a problem.

INDIVIDUALITY

The media have become masters at packaging stimuli in ways that our brains find irresistible, just as food engineers have become expert in creating "hyperpalatable" foods by manipulating levels of sugar, fat, and salt.[11] Distractibility might be regarded as the mental equivalent of obesity.

The palatability of certain kinds of mental stimulation seems to be hard-wired, just as our taste for sugar, fat, and salt is. When we inhabit a highly engineered environment, the natural world begins to seem bland and tasteless, like broccoli compared with Cheetos. Stimulation begets a need for more stimulation; without it one feels antsy, unsettled. Hungry, almost.

One consequence of this is that we are becoming more alike. I open a book of Aristotle and try to read a page of his choppy, gnomic Greek. After a few lines I start to shift my weight in the chair and drum my fingers on the table. It *is* Tuesday night, after all. I turn on *Sons of Anarchy*, and share the experience with 4.6 million of my closest friends. The next day, I have some basis for chitchat with others. I am not a freak. If I had gotten absorbed in the *Nicomachean Ethics*, my head would still be turning in a spiral of untimely meditations that could only sound strange to my acquaintances.

There is, then, a large cultural consequence to our ability to concentrate on things that aren't immediately engaging, or our lack of such ability: the persistence of intellectual diversity, or not. To insist on the importance of trained powers of concentration is to recognize that independence of thought and feeling is a fragile thing, and requires certain conditions.

What sort of ecology can preserve a robust intellectual biodiversity? We often assume that diversity is a natural upshot of free choice. Yet the market ideal of choice and attendant preoccupation with freedom tends toward a monoculture of human types: the late modern consumer self. At least the market seems to have this effect when we are constantly being addressed with hyperpalatable stimuli. What sort of outlier would you have to be, what sort of freak of self-control, to resist those well-engineered cultural marshmallows?

According to the prevailing notion, to be free means to be free to satisfy one's preferences. Preferences themselves are beyond rational scrutiny; they express the authentic core of a self whose freedom is realized when there are no encumbrances to its preference-satisfying behavior. Reason is in the service of this freedom, in a purely instrumental way; it is a person's capacity to calculate the best means to satisfy his ends. About the ends themselves we are to maintain a principled silence, out of respect for the autonomy of the individual. To do otherwise would be to risk lapsing into paternalism. Thus does liberal agnosticism about the human good line up with the market ideal of "choice." We invoke the latter as a content-free meta-good that bathes every actual choice made in the softly egalitarian, flattering light of autonomy.

This mutually reinforcing set of posits about freedom and rationality provides the basic framework for the discipline of economics, and for "liberal theory" in departments of political science. It is all wonderfully consistent, even beautiful.

But in surveying contemporary life, it is hard not to notice that this catechism doesn't describe our situation very well. Especially the bit about our preferences expressing a welling-up of the authentic self. Those preferences have become the object of social engineering, conducted not by government bureaucrats but by mind-bogglingly wealthy corporations armed with big data. To continue to insist that preferences express the sovereign self and are for that reason sacred—unavailable for rational scrutiny—is to put one's head in the sand. The resolutely individualistic understanding of freedom and rationality we have inherited from the liberal tradition disarms the critical faculties we need most in order to grapple with the large-scale societal pressures we now face.

The language of preference satisfaction and the attendant preoccupation with freedom seem ill-suited to our current circumstances, if what we want is to preserve human possibilities from going extinct. If you were to regularly air-drop Cheetos over the entire territory of a game preserve, you would probably find that *all* the herbivores preferred them right away to whatever pathetic grubs and roots they had been eating before. A few years later, the lions would have decided that hunting is not only barbaric but, worse, inconvenient. The cheetahs would come around eventually—all that running!—and the savannah would be ruled by three-toed sloths. With orange fur.

I recently visited Las Vegas, a place designed for the single purpose of separating you from your money—by tapping into your preferences. The female form is used quite freely there in advertisements, bombarding you from the moment you step off your airplane. These images work just as surely as tying a rope to a person's neck and giving it a sharp yank. Once the initial excitement wears off, you find yourself in a place that is somehow not a place. No merely local flora can compete for air and light. Nothing subtle—no feeling that isn't industrial-strength in its urgency and standardized in its appeal—can arise in such a ruthlessly monetized attentional environment.

After a day, I had to get out of there, so I rented a car. Driving through the desert, I stopped at a gas station/slot machine arcade/liquor store/fireworks emporium on an Indian reservation. A few hundred years ago, the fitness of Native Americans for the world they inhabited excited admiration in some European observers: here were natural aristocrats, disdainful of labor, dedicated to war. Unlike European peasants stooped to the grind of agriculture, anxiously accumulating grain against future want, the Indian appeared free because confident of his ability to bear hardship; leisured because tough. Whatever projections this might have involved, whatever need of the European mind was being served by the image of the noble savage, there were real cultural differences here that provided an external point of reference for self-criticism.

Then along came liquor, fast food, satellite television, methamphetamine, and all the rest. Clearly these things tapped into appetites that, before the arrival of the pertinent technologies, had been merely latent in the lifeworld of Native Americans. And clearly these candy-and-narcotics technologies played a role in their conquest and continued pacification. My impression, admittedly superficial, was that the inhabitants of this reservation were in a state of degradation that went beyond economic hardship—and that this little roadside emporium offered a glimpse into the future.

One thing that distinguishes human beings from other animals is that we are evaluative creatures. We can take a critical stance toward our own activities, and aspire to direct ourselves toward objects and projects that we judge to be more worthy than others that may be more immediately gratifying. Animals are guided by appetites that are fixed, and so are we, but we can also form a second-order desire, "a desire for a desire," when we entertain some picture of the sort of person we would *like* to be—a person who is better not because she has more self-control, but because she is moved by worthier desires.

Acquiring the tastes of a serious person is what we call education. Does it have a future? The advent of engineered, hyperpalatable mental stimuli compels us to ask the question. The transformation of the Native American lifeworld, like the transformation currently under way in our attentional environment, points up the limitations of the idea of *individual* self-determination and of exhortations to exert more self-control. We're in it together. This makes it political.

ACHIEVING A COHERENT SELF

We are wired to attend to our environment, but certain kinds of thinking require that we ignore it. Thus, when trying to recall something from memory, a person will often stare up toward the blank sky, or avert her gaze from the scene before her. Similarly, trying to predict the future and plan for it is an act of imagination that requires getting free of the present. In an influential article in *Behavioral and Brain Sciences*, Arthur M. Glenberg offers an evolutionary argument for why this kind of thinking feels effortful.

Suppressing the environment is dangerous because features of the environment that normally *should* be controlling action are ignored. "The effort is a warning signal: Take care; you are not attending to your actions!" Because it is effortful, we use suppression conservatively. Such an account makes sense of certain behaviors. Glenberg observes that "when working on a difficult intellectual problem (which should require suppression of the environment), we reduce the rate at which we are walking to avoid injury."[12]

He goes on to make the fertile suggestion that "autobiographical memory arises from suppressing the environment." Around the age of two or three years, as a child develops language, she learns to use narrative to organize and relate her experiences. By doing so, she starts to develop a coherent concept of self. This requires suppressing environmental input so the child can control what she is thinking about. And reciprocally, the ability to use language supports the ability to suppress the environment and control one's recollective experience.

While animals certainly have memory and the ability to learn, human beings are thought to be the only creatures who can deliberately recall something *not* cued by the environment.[13] But we do this only in those stretches of time when the environment is not making urgent claims on our attention. It is at these times that we try to find (or impose) coherence on our experience, retroactively. If we are currently facing a culturally and technologically induced trauma to our ability to suppress environmental input, that raises a big question: Is this distinctly human activity of coherence-finding at risk?

I think it is safe to say that our ability to suppress the environment is under greater pressure than it once was. It may be that this pressure is acutely felt only by an adult generation that developed in one attentional landscape and now finds itself inhabiting another, more highly engineered one. Younger people are famously comfortable with it all. The question remains whether we should take comfort in their comfort.

That is to say, is something important to human flourishing at risk or not? How you answer that question would seem to depend on how you understand "rational agency," to use a term of art from philosophy. Allow me to sketch two positions on this.

According to the first, what we really mean when we say that human beings tell stories and seek coherence is that we do things *for reasons*. We offer these reasons up to others (and ourselves) in language. This is what it means to be a rational agent rather than a billiard ball that is simply moved by impinging forces, or an animal that lives entirely in the moment. We have this unique tendency to want to justify ourselves, and construct a narrative that conveys the considerations that made an action seem choiceworthy. And sure, this narrative is often self-serving or self-deceptive. But however inept we may be at it, it remains true that we keep trying to "make ourselves, and our proper aspirations, articulate to ourselves," as the philosopher Talbot Brewer has written.

If Glenberg is right about memory and environmental suppression, it would seem this activity of narrative self-articulation gets under way, developmentally, with the capacity to ignore things. Further, because this self-articulation is something we are never finished with, an ability to ignore things would seem to remain important to the lifelong task of carving out and maintaining a space for rational agency for oneself, against the flux of environmental stimuli. What happens when our attention is subject to mechanized appropriation, through the pervasive use of hyperpalatable stimuli? On this first view, what is at stake in our cultural moment would seem to be the conditions for the possibility of achieving a coherent self.

But there is another position, or family of positions, that would regard this concern with a certain bemusement, because it is convinced that rational agency is an illusion. This stance is evident in a few different departments of the human sciences. Behavioral economics is impressed with psychological findings that suggest that the reasons for our actions are generally opaque to us, not objects of rational scrutiny. Whatever reason-giving we engage in tends to be a post hoc story that we tell ourselves, and is therefore beside the point if we are trying to understand human behavior. And it is indeed behavior that this discipline takes as its subject matter, not the self-understandings that accompany that behavior and give our actions their distinctly human character.

The field of neuroethics pushes this line of argument further: free will is an illusion. The experience we have of deliberating before some important decision is a mere bit of electrical chatter that our brains generate, the effect of which is to obscure from us the fact that our decision was cast before we were even aware of it. This electrical reason-chatter is said to serve some evolutionary function yet to be discovered. But regrettably, claims the neuroethicist, it also gives rise to metaphysical superstitions about the existence of *mind*.[14]

On this view, one shouldn't get too invested in making distinctions between billiard balls and human beings. And there would seem to be no reason for alarm at the transformation of our attentional landscape, as this amounts to a mere change in the array of sensory inputs impinging on the brain. The cherished "coherence" of the self is a myth we ought to grow out of anyway. We can even imagine an especially consistent neuroethicist surveying the airport scene I have described and viewing it with a certain satisfaction: maybe an environment that is sufficiently stimulating will divert us from indulging in reason-giving, that quaint activity by which man clings to the idea that he is somehow special.

Do we have to choose between this scolding antimental view and the alarm that seems warranted if we take rational agency seriously? The problem with the rationalist position as I have sketched it is that it seems *too* mental—too deliberate and individual. The rare person who has devoted himself to the examined life may consciously struggle to "make himself, and his proper aspirations, articulate to himself." But the rest of us, standing in line at the Department of Motor Vehicles? It sounds more like a midlife crisis than like something we do day-to-day.

There is another way to think about these things. What if the coherence of a life is in some significant way a function of *culture*? What if we are situated among our fellows in norms and practices that shape a life? In that case culture matters. That is, the environment matters, in a stronger way than one supposes if one adopts the interior, fully articulate model of rational agency, on the one hand, or the antimental, brain-centered view, on the other.

THE SITUATED SELF

One element of our predicament is that we engage less than we once did in everyday activities that *structure* our attention. Rituals do this, for example. They answer for us the question "What is to be done next?" and thereby relieve us of the burden of choice and reflection, as when we recite a liturgy. But I want to focus on another sort of activity, one that is neither rote like ritual, nor simply a matter of personal choice. The activities I have in mind are *skilled practices*.

Cooking an elaborate meal for an important occasion would be one example. Such practices locate the possible answers to the question "What is to be done next?" outside our own heads, in our relations to objects and to other people. They establish narrow and highly structured patterns of attention—what I shall be calling *ecologies of attention*—that can give coherence to our mental lives, however briefly. In such an ecology, the perception of a skilled practitioner is "tuned" to the features of the environment that are pertinent to effective action; extraneous information is dampened and irrelevant courses of action disappear. As a result, choice is simplified and momentum builds. Action becomes unimpeded.

In a previous book, *Shop Class as Soulcraft*, I wrote about the de-skilling of everyday life. The core theme was individual agency: the experience of seeing a direct effect of your actions in the world, and knowing that these actions are genuinely your own. I suggested that genuine agency arises not in the context of mere choices freely made (as in shopping) but rather, somewhat paradoxically, in the context of *submission* to things that have their own intractable ways, whether the thing be a musical instrument, a garden, or the building of a bridge.

A related set of ideas will be elaborated from a different angle in this book, most explicitly in Part I, "Encountering Things." There I suggest that it is indeed *things* that can serve as a kind of authority for us, by way of structuring our attention. The design of things—for example, cars and children's toys—conditions the kind of involvement we have in our own activity. Design establishes an ecology of attention that can be more or less well adapted to the requirements of skillful, unimpeded action.

The terms "submission" and "authority" are jarring to the modern ear. They may be especially unexpected here—haven't I been making a case for reclaiming our mental autonomy? But in fact, I think the experience of *attending* to something isn't easily made sense of within the prevailing Western anthropology that takes autonomy as the central human good.

Understood literally, autonomy means giving a law to oneself. The opposite of autonomy thus understood is heteronomy: being ruled by something alien to oneself. In a culture predicated on this opposition (autonomy good, heteronomy bad), it is difficult to think clearly about attention—the faculty that joins us to the world—because everything located beyond your head is regarded as a potential source of heteronomy, and therefore a threat to the self.

This sounds like an overstatement, perhaps. But it is implicit in the view of the human person we have received from certain early modern thinkers who were working out a new and quite radical notion of freedom. To do justice to the phenomenon of attention, we will have to wrestle with that notion of freedom. This is the explicit theme of the section "Interlude: A Brief History of Freedom." For now, I will simply alert the reader to be on the lookout for a somewhat paradoxical thread that runs through these pages. The paradox is that the ideal of autonomy seems to work against the development and flourishing of any rich ecology of attention—the sort in which minds may become powerful and achieve genuine independence.

In the chapters that follow we will consider the ways our environment *constitutes* the self, rather than compromises it. Attention is at the core of this constitutive or formative process. When we become competent in some particular field of practice, our perception is disciplined by that practice; we become attuned to pertinent features of a situation that would be invisible to a bystander. Through the exercise of a skill, the self that acts in the world takes on a definite shape. It comes to be in a relation of *fit* to a world it has *grasped*.

To emphasize this is to put oneself at odds with some pervasive cultural reflexes. Any quick perusal of the self-help section of a bookstore teaches that the central character in our contemporary drama is a being who must choose what he is to be, and bring about his transformation through an effort of the will. It is a heroic project of open-ended, ultimately groundless self-making. If the attentive self is in a relation of fit to a world it has apprehended, the autonomous self is in a relation of creative mastery to a world it has projected.

The latter self-understanding is an invitation to narcissism, to be sure. But it also tends to make us more easily manipulated. As atomized individuals called to create meaning for ourselves, we find ourselves the recipients of all manner of solicitude and guidance. We are offered forms of unfreedom that come slyly wrapped in autonomy talk: NO LIMITS!, as the credit card offer says. YOU'RE IN CHARGE. Autonomy talk speaks the consumerist language of preference satisfaction. Discovering your true preferences requires maximizing the number of choices you face: precisely the condition that makes for maximum dissipation of one's energies. Autonomy talk is a flattering mode of speech. It suggests that freedom is something we are entitled to, and it consists in liberation from constraints imposed by one's circumstances.

The image of human excellence I would like to offer as a counterweight to freedom thus understood is that of a powerful, independent mind working at full song. Such independence is won through disciplined attention, in the kind of action that joins us to the world. And—this is important—it is precisely those constraining circumstances that provide the discipline.

This claim—about the role of attention in bringing the self into a relation of fit to the external world—is part of a broader anthropological assertion that runs through the book: we find ourselves *situated* in a world that is not of our making, and this "situatedness" is fundamental to what a human being is.

I will be emphasizing three elements of this situatedness: our embodiment, our deeply social nature, and the fact that we live in a particular historical moment. These correspond to the three major divisions of the book: "Encountering Things," "Other People," and "Inheritance." In these divisions I will reinterpret what are often taken to be encumbrances to the personal will in the modern tradition—sources of unfreedom—and identify them rather as the framing conditions for any worthwhile human performance.

It would be conventional at this point to say that what emerges in the argument is a concept of true freedom as opposed to false freedom. What I want to do instead is simply drop "freedom" as a term of approbation. The word is strained by being made to do too much cultural work; it has become a linguistic reflex that affirms our image of ourselves as autonomous. In doing so, it obscures the sources of our current predicament of attention—by reenacting the central dogma that gave rise to it.

For several hundred years now, the ideal self of the West has been striving to secure its freedom by rendering the external world fully pliable to its will. For the originators of modern thought, this was to be accomplished by treating objects as projections of the mind; we make contact with them only through our *representations* of them. Early in the twenty-first century, our daily lives are saturated with representations; we have come to resemble the human person as posited in Enlightenment thought. Such is the power and ubiquity of these representations that we find ourselves living a highly mediated existence. The thing is, in this style of existence we *ourselves* have been rendered pliable—to whoever has the power to craft the most bewitching representations or to control the portals of public space through which we must pass to conduct the business of life.

Autonomy talk stems from Enlightenment epistemology and moral theory, which did important polemical work in their day against various forms of coercion. Times have changed. The philosophical project of this book is to *reclaim the real*, as against representations. That is why the central term of approbation in these pages is not "freedom" but "agency." For it is when we are engaged in a skilled practice that the world shows up for us as having a reality of its own, independent of the self. Reciprocally, the self comes into view as being *in a situation* that is not of its own making. The Latin root of our English word "attention" is *tenere*, which means to stretch or make tense. External objects provide an attachment point for the mind; they pull us out of ourselves. It is in the encounter between the self and the brute alien otherness of the real that beautiful things become possible: the puck-handling finesse of the hockey player, for example.

Encountering the world *as* real can be a source of pleasure—indeed of quasi-religious feelings of wonder and gratitude—in light of which manufactured realities are revealed as pale counterfeits, and lose some of their grip on us. It is not that in becoming skilled one somehow becomes immune to distraction. I do believe this book has therapeutic implications, but they are not so immediately obvious as that. Rather, the cultural crisis of attention provides an occasion to examine the big anthropological picture we have been operating within since the Enlightenment, and to revisit the question of how we stand in relation to the world beyond our heads. Anything less far-reaching would be inadequate to the challenges we face.

PART I

ENCOUNTERING THINGS

1

THE JIG, THE NUDGE, AND LOCAL ECOLOGY

When a carpenter wants to cut a half-dozen boards to the same length, he is unlikely to measure each one, mark it, and then carefully guide his saw along the line he has made on each board. Rather, he will make a jig. A jig is a device or procedure that guides a repeated action by constraining the environment in such a way as to make the action go smoothly, the same each time, without his having to think about it. If he is on a job site rather than in a workshop, he will make the jig out of whatever is on hand, maybe using the clean line of a freshly laid cinder-block wall as a stop to butt each board up against, side by side on sawhorses. He'll make a measurement on the first and last board, maybe snap a chalk line across the marks, then tack a straight piece of scrap plywood along that line, traversing the whole array of boards, to serve as a guide for his saw. Then he just has to slide his saw along the plywood edge and presto, six boards of perfectly equal length.

A jig reduces the degrees of freedom that are afforded by the environment. It stabilizes a process, and in doing so lightens the burden of care—on both memory and fine muscular control. The concept of a jig can be extended beyond its original context of manual fabrication. As David Kirsh points out in his classic and indispensable article "The Intelligent Use of Space," jigging is something that expert practitioners do generally, if we allow that it is possible to jig one's environment "informationally."

A bartender gets an order from a waitress: a vodka and soda, a glass of house red, a martini up, and a mojito. What does he do? He lays out the four different kinds of glass that the drinks require in a row, so he doesn't have to remember them. If another order comes in while he is working on the first, he lays out more glasses. In this way, the sequence of orders, as well as the content of each order, is represented in a spatial arrangement that is visible at a glance. It is in the world, rather than in his head. This is good, because there is only so much room in his head.

Consider a short-order cook on the breakfast shift. As he finishes his coffee, the first order of the morning comes in: a sausage, onion, and mushroom omelet with wheat toast. The cook lays out the already chopped sausage next to the pan, the onions next to the sausage, then the bread, and finally the mushrooms, farthest from the pan. He now has the ingredients in a spatial order that corresponds to the temporal order in which he will require them: once it gets hot, the sausage will provide the grease in which the onions will cook, and the onions take longer to fry than the mushrooms do. He places the bread between the onions and the mushrooms as a reminder to himself to start toasting the bread at such a time that the toast will be ready just as he is sliding the omelet out of the pan. The pace of what comes next is set by the level of heat under the pan, which he generally leaves at the same level throughout the shift—it corresponds to an internal clock he has developed through long practice. When the sound and smell of the omelet indicate that he ought to turn down the heat, he removes the pan from the flame and sets it to the side for a while—maybe the amount of time it takes to retrieve a colander—rather than turn down the flame. That way, the level of heat is encoded spatially in the environment, in a way accessible to peripheral vision, and has a temporal dimension too, becoming part of the cook's bodily rhythms as he moves around the kitchen. He doesn't have to stoop down to look at the flame and make fine adjustments to a knob. The mental work he has to do on this omelet is reduced and externalized in the arrangement of physical space.

Kirsh finds that experts "constantly re-arrange items to make it easy to 1. track the task; 2. figure out, remember, or notice the properties signaling what to do next; 3. predict the effects of actions." He has observed cooks leaving a knife or other utensil next to the ingredient to be used next, serving to mark its place in the action plan. This frees them from the kind of halting deliberation that you can see at a glance in the movements of a beginner who is relying on conscious analytical processes. Experts make things easier for themselves by "partially jigging or informationally structuring the environment as they go along."[1]

A physical jig reduces the physical degrees of freedom a person must contend with. By seeding the environment with attention-getting objects (such as a knife left in a certain spot) or arranging the environment to keep attention *away* from something (as, for example, when a dieter keeps certain foods out of easy view), a person can informationally jig it to constrain his mental degrees of freedom. The upshot is that to keep action on track, according to some guiding purpose, one has to keep attention properly directed. To do this, it helps a great deal to arrange the environment accordingly, and in fact this is what is generally done by someone engaged in a skilled activity. Once we have achieved competence in the skill, we don't routinely rely on our powers of concentration and self-regulation—those higher-level "executive" functions that are easily exhausted. Rather, we find ways to recruit our surroundings for the sake of achieving our purposes with a minimum expenditure of these scarce mental resources.

High-level performance is then to some degree a matter of being well situated, let us say. When we watch a cook who is hitting his flow, we see someone *inhabiting* the kitchen—a space for action that has in some sense become an extension of himself.

As orders pile up and overlap, the available work space in the kitchen cannot remain devoted to separate orders, with ingredients arranged to match a definite temporal sequence. It becomes messy-looking to a casual observer, and necessarily improvisational because the cook is dealing with competing structures of sequence: the sequence of orders received; the sequence that might be more efficient by grouping orders requiring the same task, or tasks done in close proximity to each other; the sequences that arise from the fact that different amounts of time are required to cook different kinds of food; and of course the desired outcome that all orders of a given party arrive simultaneously, good and hot.

Maybe there is a new prep cook who sliced the green peppers in thicker slices last night, so they take longer to cook. For the cook who is on fire, jacked up on an awareness of his own full-firing improvisational chops, this hitch does nothing but add a little syncopation to his internal cooking clock. He spins on his heel, does a little *I, Robot* dance move, and seamlessly hits upon a task that fits into the extra, unanticipated forty-five seconds it takes to get those peppers soft enough to add to the omelet that is just now skinning over. "I'm a *machine*!" He lets the servers know it. The busier it gets, the more "on" he is.

Such moments probably don't arise in a push-button McDonald's kitchen, modeled on the assembly line. In such a setting the jig is very elaborate, and rigidly deployed by someone other than the worker him- or herself. The point of an assembly line is to replace skilled work with routinized work that can be done by unskilled labor. Early in the twentieth century this gave rise to the saying "Cheap men need expensive jigs; expensive men need only the tools in their toolbox."

The jig as it is used in a skilled practice is located somewhere between the overdetermination of the assembly line and the ideal of autonomy. In the tension between freedom and structure, which shows itself with special clarity in skilled practices, there is something important to be learned about human agency in general.

A humming kitchen of the sort I have described may be regarded as an ecology of attention in which the external demand of feeding people in a timely manner provides a loose structure within which the kitchen staff themselves establish an internal order of smooth, adaptive action. In the course of doing this they hit upon various jigs for keeping their attention properly directed.

This is consistent with a shift currently taking place at the frontiers of cognitive science, in the (still somewhat dissident) movement toward a picture of human beings as having "extended" or "embedded" cognition. Andy Clark, one of the leading figures in the extended-mind literature, writes that "advanced cognition depends crucially on our ability to *dissipate* reasoning: to diffuse achieved knowledge and practical wisdom through complex structures, and to reduce the loads on individual brains by locating those brains in complex webs of linguistic, social, political and institutional constraints."[2] Such constraints might be called cultural jigs.

Consider an obvious example of how our capacity for "advanced cognition" depends on environmental props: doing arithmetic. It is not hard to multiply 18 by 12 in your head, for example by multiplying 18 by 10 to get 180, and then multiplying 18 by 2 to get 36, and finally adding 36 to 180 to get 216. We break the problem down into simpler pieces, to be reassembled at the end. We can do this because our "working memory" is able to keep three to five items in play at any one time. But no more than that, for most of us (this is one of the more robust findings in cognitive science).[3] If one has to multiply 356 by 911, the number of items to juggle becomes quite challenging, so what do we do? We reach for a pencil and paper.

With this simple expedient, we vastly extend our intellectual capacities: long division, algebra, calculating the load on a structural member, building space shuttles, and all the rest. The reader may have had the experience of being unable to think without a pen in hand, or a laptop open. A number of metaphors have been suggested: we "offload" some of our thinking onto our surroundings, or we incorporate objects in such a way that they come to act like prosthetics. The point is that to understand human cognition, it is a mistake to focus only on what goes on inside the skull, because our abilities are highly "scaffolded" by environmental props—by technologies and cultural practices, which become an integral part of our cognitive system.[4]

Could this same argument be applied to our moral capacities? We have already touched upon the idea that there is no clean division to be made between the narrowly "cognitive" capacity for mental concentration and the moral capacity for self-regulation. Let us ease our way into this question. We can begin by taking our bearings from a contemporary quarrel in the world of public policy, and see if there is anything interesting to be said about it from the perspective of the situated self.

THE NUDGE

The view of human beings that prevailed in economics and public policy in the twentieth century seems implausible in retrospect: it held that we are rational beings who gather all the information pertinent to our situation, calculate the best means to given ends, and then go about optimizing our choices accordingly. The assumption was that we are able to do this because we know what we want, and the calculation will be simple because our interests are not in conflict with one another; each can be located on the same "utility" scale, which has only one dimension.

This "rational optimizer" view has come in for thorough revision with the advent of the more psychologically informed school of "behavioral economics." There is a large literature that shows that, for example, we consistently underestimate how long it will take us to get things done, no matter how many times we have been surprised by this same fact in the past (the so-called planning fallacy). We give undue weight to the most recent events when trying to grasp a larger pattern and predict the future. In general, we are terrible at estimating probabilities. We are not so much rational optimizers as creatures who rely on biases and crude heuristics for making important decisions.

In *Nudge*, Cass Sunstein, the former head of the Office of Information and Regulatory Affairs under President Obama, and the economist Richard Thaler argue for a mode of social engineering that takes account of these psychological facts.[5] For starters, we're a lot lazier than the rational optimizer view would have it. That is, to make everything a matter for reflection and explicit evaluation goes against the grain of how human beings normally operate. So, for example, if one wants to increase the savings rate, it makes a great deal of difference whether employers set the default so that employees have to opt in to a 401(k) plan if they want it, or instead they have to opt out if they don't. Participation is much higher under opt-out. In general, when we are faced with an array of choices, how we choose depends very much on how those choices are presented to us (to the point that we will choose against our own best interests if the framing nudges us that way). Here, then, is an opportunity for a fairly unobtrusive bit of social engineering that doesn't force anyone to do anything; it just steers us in one direction rather than another.

We might call this an administrative jig. But note that this kind of administering of human beings, which certainly has its place in a modern state, is quite different from the jig as it appears in skilled practices. The difference is that skilled practitioners *themselves* keep their actions on track by "partially jigging or informationally structuring the environment *as they go along*," as Kirsh says (emphasis added). The jig itself is not flexible—indeed being rigid is the whole point of a jig—but it is deployed flexibly in the intelligent ordering of the environment by someone who is in command of his own actions. The local, actor-centered use of the jig is more attractive, to my mind, than the prospect of being nudged by Cass Sunstein.

Let's note right away that there is a risk of misstating the contrast between the jig and the nudge by putting too much emphasis on the jig being a *creation* of the agent himself. Quite apart from the extreme case of the push-button McDonald's kitchen, it is true in general that a cook begins his day in an environment that has already been given a long-term structure by someone else, equipped with tools and facilities laid out in some arrangement. This might be called the background jig. A further part of the background jig is the menu: only certain dishes may be ordered. That is, the menu regulates the cook's activity. And the prep work (chopping vegetables, preboiling the potatoes for home fries, etc.) has been done by the evening shift, who are now in bed. Thus, *other people* tacitly hover in the background of the cook's activity and give shape to it.

The ideal of autonomy therefore doesn't capture what we are interested in when we recognize that there is something valuable going on in the kitchen and want to understand it by way of contrast to the nudge. For the ideal of autonomy is built around the notion of a sovereign self, whose sovereignty consists in having everything within full view, available to her as material for her own choice, planning, and optimization. There are no determinants of her actions that she doesn't have a handle on. This picture doesn't comfortably admit our dependence on others, or the ways our freedom is ordered by various framing conditions we have inherited, which are not of our own making.

The contrast I want to make between the jig and the nudge thus lies elsewhere; it is not a brief for autonomy. Rather, what is at issue is the source of external authority: administrative fiat or something more organic, deriving from the social world.

CULTURAL JIGS

Consider once again the problem of saving money, a favorite example of the nudgers. The imperative to be thrifty was once part of a larger cultural setting: the Protestant ethic, famously explained by Max Weber. To accumulate wealth was important not as a means to indulgence, but as a sign that one's life was on track. God had so arranged things that the status of one's soul was visible in one's portfolio; wealth was proof of election.

And even apart from such supernatural props, there was in early capitalism a perfectly this-worldly discredit that fell on the spendthrift. "Be frugal and free," said Benjamin Franklin. The republican personality took pride in his freedom, and was wary of any debt that would compromise it. The debtor cannot speak frankly to the man he owes money to; he must make himself pleasing and hope for continued forbearance. Yet frank speech, or "free speech," is the basis for specifically democratic social relations; the democrat's pride lies in not being a flatterer of any man.

There was, then, a cultural jig supporting thrift that was richly elaborated in the dominant religion and in our political psychology; a nestled set of mutually reinforcing moral norms gave a certain shape to life in early America (which is not to say that this was the best possible shape). The invention of consumer credit early in the twentieth century did a fair bit to dismantle this jig. The historian Jackson Lears has explained how, with this innovation, previously unthinkable acquisitions became thinkable through the installment plan, and more than thinkable: it became normal to carry debt.

Here is the point. Getting people to save money through

administrative nudges such as the opt-out 401(k) plan is best seen not as a remedy for our failure to be rational *as individuals*, but as an attempt to compensate for the dismantling of those cultural jigs we once relied on to act (and think and feel) in ways that support thrift. The norms that cultural jigs express and reinforce tend to be reiterated, fractal-like, along different axes of social life; they are robust in that way. Together they make up a more or less coherent form of ethical life, for example Protestant republicanism. By contrast, administrative nudges are a thin attempt to get us to act *as if* we were virtuous, without any reference to character traits like self-control.

But wait. Is the Protestant's virtue "situational"? If we plucked him out of eighteenth-century New England and set him down in Tahiti, would we discover a different man? Such that we would have to say that his self-control is precisely *not* a deep attribute of his person, as we normally take character to be? And if so, on what basis can one prefer Calvin to Cass Sunstein? One way to parse this is to think about habit and formation. The word "character" comes from a Greek word that means "stamp." Character, in the original view, is something that is stamped upon you by experience, and your history of responding to various kinds of experience, not the welling up of an innate quality. Character is a kind of jig that is built up through habit, becoming a reliable pattern of responses to a variety of situations. There are limits, of course. Character is "tested," and may fail. In some circumstances, a person's behavior may be "out of character." But still, there is something we call character. Habit seems to work from the outside in; from behavior to personality. One question, then, is whether an administrative nudge, which works on behavior, could have an effect similar to a cultural jig such as Protestantism. Both help to regulate life. But there is a big difference in how this regulation operates. If I fail to opt out of a 401(k), have I really acted? Have I *done* something, such as facing down temptation, that helps to wear the groove of habit into my character? Probably not.[6]

It is tempting, then, to criticize the nudge from the perspective of communitarianism—I mean the Burkean sort that seeks to preserve the rich soil of historically well-sedimented norms and practices. I find this perspective attractive. It is valuable as a point of orientation on our critical compass. But to bring it to bear against the nudge argument would be to misidentify the field of forces into which that argument seeks to intervene, and thereby to miss its critical force. Thaler and Sunstein are not enlighteners seeking to expand the empire of rational administration by chasing away settled forms of social authority. Rather, they are drawing attention to the fact that we are already administered in various ways, inescapably, but are generally not aware of it. And this has everything to do with the managing of our attention by others.

Consider the supermarket. The placement of items on the shelves is not haphazard, but the result of a negotiation in which companies compete for prime real estate: at eye level, or at the slow-moving checkout line. This is an example of how our decisions take place in an environment that has been given shape by "choice architects," as Thaler and Sunstein say. And this is inevitable. The shelves will necessarily have *some* arrangement. The only question is whether this arrangement is determined by a simple auction for our attention, to be won by the highest bidder, or might be subject to a more public-spirited calculus and made to serve the interests of consumers themselves. In the latter case, having Frosted Flakes at child's-eye level is probably not what we would prescribe. Do you want to be nudged by the Office of Information and Regulatory Affairs or by Kellogg's? Sunstein presumably doesn't get any personal financial gain by putting Brussels sprouts within easy reach (though we can't rule out the possibility that he gets some perverse bureaucratic pleasure).

If you shop at the supermarket (because you aren't Amish), the alternative to being nudged by the choice architects of a central political authority isn't to inhabit some well-ordered local community. It is to be manipulated by *other* choice architects, equally distant, who act on behalf of their shareholders without any accountability to the common good. Two cheers for the nudge, then.

A third cheer would have to be somewhat ironic. The studies that inform behavioral economics investigate an individual in the artificial setting of a university psychology lab, where the whole point is to isolate and control every variable. But this means she has been denuded of any environmental props she may rely on in everyday life. From the perspective of the extended-mind literature, it is not surprising that these studies show that we are poor reasoners *in isolation*. The fact that this artificial person has so little skill in practical reasoning is what authorizes the nudgers' spirit of supervision.

The ironic third cheer comes from recognizing that this beginning premise of the solitary reasoner is false, and yet apt for describing us in certain situations. When we are in the supermarket or any idealized consumer space (for example, alone with one's laptop and a credit card), don't we in fact resemble the isolated subject of a psych experiment? As such, we are ideal raw material for the architects of mass behavior, and we do well to be aware of the fact so we can choose our architect.

JIGS FOR HIRE

In the boom after World War II, the left lost interest in economics and shifted its focus from labor issues to a more wide-ranging project of liberation, to be achieved by unmasking and discrediting various forms of cultural authority. In retrospect, this seems to have prepared the way for a new right, no less committed to the ideal of the unencumbered self (that ideal actor of the free market), whose freedom could be realized only in a public space cleared of distorting influences—through deregulation.

Few institutions or sites of moral authority were left untouched by the left's critiques. Parents, teachers, priests, elected officials—there was little that seemed defensible. Looking around in stunned silence, left and right eventually discovered common ground: a neoliberal consensus in which we have agreed to let the market quietly work its solvent action on all impediments to the natural chooser within.

Another way to put this is that the left's project of liberation led us to dismantle inherited cultural jigs that once imposed a certain coherence (for better and worse) on individual lives. This created a vacuum of cultural authority that has been filled, opportunistically, with attentional landscapes that get installed by whatever "choice architect" brings the most energy to the task—usually because it sees the profit potential.

The combined effect of these liberating and deregulating efforts of the right and left has been to ratchet up the burden of *self*-regulation.[7] Some indication of how well we are bearing this burden can be found in the fact that we are now very fat, very much in debt, and very prone to divorce.

The effects of this have not been evenly distributed. To gain admission to the svelte, solvent middle class, and stay there, now requires extraordinary self-discipline. Such discipline is generally inculcated in families. Two self-disciplined people meet in graduate school, mate, and pass their disciplined ways on to their children. But we also make use of external props that are available to those with means: jigs for hire.

I know something of this firsthand. I went seven or eight years without filing a tax return, not out of a determination to avoid taxes, but simply due to the soul-crushing paralysis that would come over me at the thought of facing an impenetrable morass of tax instructions, and my own negligence in keeping records. Worried about going to jail, I eventually faced the music and, through a colossal effort, set things right with the IRS.

I am wealthier now than I was then, and this has allowed me to hire an accountant. I have off-loaded the burden of tax compliance onto her completely—not just the cognitive burden of it, but the disciplinary burden of staying on top of deadlines. I pay her to nag me. To do what? Merely to stop by her office once per quarter and sign some stuff, which I hardly even look at. I totally love this arrangement.

The point is that staying out of jail is *much* easier with money. The daunting complexity of the demands we have to comply with, and the opportunities for diversion that abound, add up to a pretty compelling case for just staying on the couch in a state of overstimulated avoidance. If you have ever sat in a municipal court waiting your turn to go before the judge, listening to other people's woes, you know that "failure to appear" is the most common rap, and people are led off in handcuffs for this every day, the last step in a concatenation of fuckups that may have begun with their failure to complete some bureaucratic task.

We all know how easy it is for a life to go off track when we are left to our own devices, and that is why those of us with means do what we can to jig the way for our children. I worked for one of the test-prep companies for about six months, coaching students for the SAT and GRE tests. The intellectual content of what I was offering was pretty close to zero—a few tips that could be put on one side of an index card. But the classes and tutoring sessions provided an institutional setting that forced students to show up and do practice tests. The benefit was mostly one of providing a jig for hire that helped relieve students of the burden of self-regulation. It also helped to relieve parents of the burden of discipline. Parental authority was a central target of the sixties counterculture. Now we want to be our children's best friends, and this is easier if you can outsource the discipline. Disciplining children is one of the most thankless tasks in a marriage, and a persistent source of resentment between many spouses. Thank God for professional help—especially if the legal and cultural barriers to divorce are low.

With help such as the kind I was offering at the test prep center, students eventually find their way to graduate school, pair up, and reproduce. What is being reproduced is social capital—all those capacities, habits, relationships, and institutional certifications that a person needs to thrive. Such social capital seems to be more tightly correlated with money capital than ever before. Maybe one reason for this is that the cultural jigs once relied on by the middle class have been widely dismantled, in the name of personal autonomy. (For example, the marital jig was weakened by the advent of no-fault divorce in the 1970s.) The costs and benefits of such autonomy don't always accrue to the same parts of society, and I think that is because the disciplinary functions of culture have in fact not been dissolved so much as privatized. They are located less in a shared order of meaning such as Protestant thrift, parental authority, or injunctions against gluttony, and more in the professional nagging services provided by financial planners, tutors, and personal trainers. The blessings of personal autonomy have been expressed with most eloquence, whether of the libertarian or the lefty kind, by opinion-makers who probably have the means to avail themselves of such services.

But let us return to our cook. What he seems to convey when he exults "I'm a *machine*," ironically enough, is that he is caught up in a moment of savoring his own distinctly human excellence. One virtue of the extended-mind rubric is that it offers a theoretical frame for understanding a very basic, low-to-the-ground mode of human flourishing, in which we are wholly absorbed in activity that joins us to the world and to others. The cook finds pleasure in his ability to improvise; to meet the unpredictable demands of the situation, and to do so within the structure imposed by the kitchen. Does this narrow example shed any wider critical light on the liberationist project to dismantle shared cultural jigs? Living skillfully requires that some things be settled. At the least, the example of the cook should alert us that the ideal of freedom from external influence doesn't capture all the elements that contribute to an impressive human performance. But at this stage in our argument it would be a strain to draw large cultural conclusions from this. We need to consider more fully the kind of cognitive extension that happens when we become skilled.

2

EMBODIED PERCEPTION

When a camera caught the Pittsburgh Penguins center Tyler Kennedy licking his hockey stick after a shift against the Florida Panthers, the video went viral on YouTube. The reaction of hockey players to the video was somewhat different from the simple disgust expressed by the general public: to be sure, licking your stick isn't something you really want to be seen doing in public, but the impulse is perfectly understandable.

Among the bonds that athletes form with their equipment, that between an ice hockey player and his stick is perhaps the most intimate. He holds it in his hands roughly five times as many hours per week during the ice hockey season as the average major league baseball player holds his bat, including batting practice. It becomes very much an extension of the player's body, as David Fleming explains in "To Each His Own," his excellent account in *ESPN Magazine* that focuses on the NHL's leading scorer in 2007–2008, Alex Ovechkin. Fleming notes that in addition to using his stick to pass and shoot, Ovechkin "uses it as a crutch to get up after big hits. He wields it high and with two hands, like a nightstick, in scrums in front of the net . . . He taps it against the boards to applaud a teammate after a fight or against the goaltender's shin pads after a good save. He uses his stick to open and close the bench door. To calm his nerves late in games, Ovechkin will sit on the boards with his back to the ice and his stick in his lap, like a baby blanket, and lovingly retape the blade." While moving at high speeds in a hostile field of overgrown bodies and sharp skates, an ice hockey player must handle a small object from the far end of a long stick—an object that is prone to both sliding and rolling. It is a game at once of violence and of finesse.

Like an amputee who modifies and fine-tunes his prosthetic, the Washington Capitals coach Bruce Boudreau recalls prepping his stick for games during his days as a player. He would "sit in his kitchen and customize the fiberglass curve of his weapon by carefully steaming it over a teakettle. Then he'd wedge it under a door hinge and bend it until it was perfect, race outside and plunge it into the snow to set the blade. With the kettle at a boil, he'd have a cup of tea while waiting for the snow to complete its work." The knob of tape he would make on the end of his sticks had his own signature to it. Handed an old stick by a fan, Boudreau instantly recognized it as his own, from thirty-five years earlier. "Our sticks become part of our DNA," he says. (The fan took the stick back and said he'd return it if the Capitals won the Stanley Cup.)

There is a very real sense in which a tool may be integrated into one's body, for one who has become expert in using the tool. There is a growing number of studies that support this idea of "cognitive extension"; the new capacities added by tools and prosthetics become indistinguishable from those of the natural human body, in terms of how they are treated by the brain that organizes our actions and perceptions.[1]

What is it like to be Alex Ovechkin? Perhaps we can regard his case as a more highly developed version of the excellence we were trying to understand in the case of the short-order cook. In a game like hockey, it is the rules of the game that make up the jig. It is played with regulation-sized sticks, on a regulation-sized rink, but within these parameters the scope for individual finesse seems to be inexhaustible. There is pleasure in those moments when we feel a growing mastery in some small domain—mastery that is exercised through our bodies, with the use of tools. To understand this, let us consider the way our attention is structured in a skilled practice, such that a tool may be incorporated into one's body.

Consider the experience of using a probe to explore an unseen space, or the way a blind person feels his way by tapping with a stick. At first you feel the varying pressure of the probe against your palm and fingers, and you have to *interpret* this pressure, mapping it in some as yet uncertain way onto a spatial representation that you are developing of the object. But as you learn to use the probe, your awareness of this pressure at the handle end is transformed into something quite different. What you have eventually is a direct, unmediated sense of the probe's tip touching the objects you are exploring. You are no longer attending to the sensations in your hand.

The philosopher Michael Polanyi analyzed this moment when we achieve competence with a probe, and in doing so found that he had to use the word "attend" in a new formulation: you are now "attending *from*" the sensations in your hand *to* the objects at the probe's tip; the sensations themselves you are only "subsidiarily aware of." In this way "an interpretive effort transposes meaningless feelings into meaningful ones, and places these at some distance [i.e., the length of the probe] from the original ones. We become aware of the feelings in our hands *in terms of* their meaning . . . to which we are attending." This meaning is located at the tip of the probe. The probe itself becomes transparent; it disappears. You are no longer engaged in an interpretive effort. The crucial fact that makes this integration of a prosthetic possible is that there is a *closed loop* between action and perception: what you perceive is determined by what you do, just as when we make use of our own hands. Let us go a little deeper into the mutual entanglement of action and perception.

EMBODIED PERCEPTION

Consider a blossoming crape myrtle tree that I see in my backyard. How does it present itself? What I actually *see* is one side, one profile, of what I take to be a three-dimensional tree. Though I have never paid it much mind until now, I suppose I have seen the back side of the tree many times in the past, and I can make a point of going around and looking at it later, but while I am looking at the tree now its back side is not *perceptually* given. Yet my apprehension of the tree as a three-dimensional whole has an immediate character to it. It doesn't feel like a recollection, or a consideration of hypothetical possibilities in the future. I "see" (in some sense that we need to investigate) the whole tree at once. In everyday life (as opposed to a freshman philosophy class), I certainly don't worry that the back side of the tree is *not there* when I can't see it.[2]

These facts of experience present something of a puzzle if we understand perception simply as the brain's response to stimulation. To preserve that view, one has to posit some mechanism by which the discrete perceptual profiles of the tree somehow get integrated into a whole by the brain; there must be some kind of "processing" that leads to a "representation" of the tree in the mind.[3] The basic supposition in this, the standard view, is that vision may be understood by analogy with a still photograph. The more rounded-out sense we have of the world is then taken to be the product of a dynamic 3-D modeling that our brains do, just like the software used by animation studios.

An alternative to this approach begins from a different set of facts: our brains are connected to eyes that are free to move within their sockets, located in a head that swivels on a neck, attached to a body that moves around on the ground, in ways that are characteristic of the particular sort of bipedal animals that we are. According to a school of thought that has been gaining traction in the last fifteen years, these facts—our embodiment, and the possibility of movement that our bodies provide—are no mere accessory to perception, but rather constitutive of the way we perceive. As one researcher puts it, "Perceiving is a way of acting. Perception is not something that happens to us, or in us. It is something we do."[4]

James J. Gibson spent his early career as a psychologist assessing the aptitude of candidates for pilot training during World War II. Over the course of several decades' research on visual perception, he began to chafe against the fundamental assumption that "sensory inputs are converted into perceptions by operations of the mind." His 1979 book *The Ecological Approach to Visual Perception* announced a subtle but important reconception of vision, not as the purely mental processing of sensory inputs, but rather as an activity in which we use our body to "extract invariants from the stimulus flux." In other words, one has to be able to explore a scene from different perspectives to perceive what remains the same about it—its nature and structure, regardless of perspective—and locomotion is an indispensable part of this process.[5]

Further, there is evidence that only *self*-motion accomplishes this; the visual system cannot develop if one is merely transported around, passively. In one of the earliest experiments in what would come to be called embodied cognition, ten pairs of kittens were reared in the dark, except for three hours per day that they spent in a carousel apparatus that allowed one twin to move freely, while the other was carried passively by the movements of the first. The active kitten could move up, down, away from, or toward the center of the carousel, as well as rotate in epicycles at the periphery of the carousel's radius. The kittens could not see each other, and the surrounding environment was contrived so that both kittens received identical visual stimulation as they moved around; the only difference was that one moved itself, the other was passively carried. The active kittens developed normally; the passive kittens failed to develop visually guided paw placement, avoidance of a visual cliff, a blink response to quickly approaching objects, or visual pursuit of a moving object.

The still photograph turns out to be a poor metaphor for understanding visual perception, for the simple reason that the world is not still, nor are we in relation to it.[6] This has far-reaching consequences, because some foundational concepts of standard cognitive psychology are predicated on the assumption that we can understand the eye by analogy with a camera, in isolation from the rest of the body. Nor is this a mere intramural fight between quarreling academic camps; what is at issue is the question of how we make contact with the world beyond our heads.

In the domain of visual perception, cognitive psychology set out to solve a certain puzzle: an indefinite variety of three-dimensional objects can project identical two-dimensional shapes on the retina of an observer. If static optical information is all that is available to the subject, then because such information underspecifies the shapes of surfaces, it follows that it must be supplemented with something else; something going on inside the head of the subject—namely, assumptions about the structure of the world. This is the motivation for thinking that perception involves an *inferential* process in the brain. This inference is taken to be computational. That is, "cognition consists in the manipulation of symbols, where these manipulations often involve the application of rules for the purpose of deriving conclusions that go beyond the information [that is presented to the eye]," as Lawrence Shapiro writes in *Embodied Cognition*, his excellent overview of the embodied cognition literature.[7]

A single retinal image is certainly not adequate to the task of specifying the world, but the visual stimulus received over time by an observer in motion *is* adequate, Gibson argues, and so on his account *the whole motivation for conceiving perception as involving inference and computation collapses.* This is completely revolutionary.[8] The brain does not have to construct a representation of the world. The world is known to us because we live and act in it, and accumulate experience.

Surprisingly, it is in the field of robotics that some of the most convincing evidence has emerged that inference, calculation, and representation are a grossly inefficient way to go about negotiating a physical environment. In his now-classic article "Intelligence Without Representation," published in the journal *Artificial Intelligence* in 1991, Rodney Brooks wrote that "the world is its own best model." Roboticists are learning a lesson that evolution learned long ago, namely, that the task of solving problems needn't be accomplished solely by the brain, but can be distributed among the brain, the body, and the world.

Consider the problem of catching a fly ball. According to the standard view, we might suppose that the visual system provides inputs about the current position of the ball, and a separate processor (the brain) predicts its future trajectory. How we might do this is a bit mysterious, given that most of us wouldn't be able to calculate such a trajectory consciously, with pencil and paper. The Gibsonian approach suggests we don't need to do any such thing, whether consciously or subconsciously. And in fact what we do, it turns out, is run in such a way that the image of the ball appears to move in a straight line, at constant speed, against the visual background.[9] It so happens that finding and exploiting *this* invariant, which is *available* in the optic flow *if* you run just right, puts you in the right spot to catch the ball. (The same strategy appears to be used by dogs who catch Frisbees, even on windy days.) You don't need an inner model of the pseudo-parabolic trajectories that baseballs follow, with corrections for air resistance at different altitudes and so forth. It's a good thing, too.

We think through the body. The fundamental contribution of this school of psychological research is that it puts the mind back in the world, where it belongs, after several centuries of being locked within our heads. The boundary of our cognitive processes cannot be cleanly drawn at the outer surface of our skulls, or indeed of our bodies more generally. They are, in a sense, distributed in the world that we act in.

A good way to capture the force of this conceptual shift is to compare a humanoid robot designed on the principles of "good old-fashioned artificial intelligence" with a robot that reflects the new ecological thinking. Andy Clark has done just this (and much more) in his hugely illuminating book *Supersizing the Mind.*

Walking may be understood as a kind of problem solving. Asimo, a biped robot built by Honda, relies on precise control of its joint angles by motors, servos, and other mechanical actuators, which enables it to negotiate such challenges as a flight of stairs. The amount of energy Asimo requires to shift a given unit of weight a given distance is about sixteen times greater than that which a human body requires. Commenting on this general approach to building robots, one roboticist remarked that a specimen like this isn't very pleasing to watch; it suffers from "a kind of rigor mortis."

The efficiency of a human body has been matched, however, by a very different kind of contraption, one that relies on its own "passive dynamics." A passive-dynamic walker is powered only by gravity, somewhat like the old toy Slinky walking down the stairs. A walking robot with legs and knees and swinging arms developed in 2001 requires only a slight downward slope. It walks smoothly, with an uncannily human gait. There is no control system. Its movement is not the result of movement planning and calculation, based on some representation of the world, but is rather a function of its form: the lengths of its limbs, their weights, and the damping and spring rates of the joints connecting them, much as muscles and ligaments connect the limbs of a human body. A powered robot can exploit the same design principles.

This is an instance of "ecological control," or "morphological computation," in which "goals are not achieved by micromanaging every detail of the desired action or response but by making the most of robust, reliable sources of relevant order in the bodily or worldly environment of the controller," as Clark writes. The "processing," as it were, is partially taken over by the dynamics that are inherent in the interaction between the robot and its environment.

A toddler begins to walk by learning to exploit the passive dynamics of his own body. Initially his body (which of course is growing and changing) is experienced as a beginner experiences a hockey stick; it is obtrusive and frustrating. The infant learns through exploration "which neural commands bring about which bodily effects," and with enough practice he becomes "skilled enough to issue those commands without conscious effort."[10] At that point the child's body has become transparent in the same sense that a blind man's probe becomes transparent; it disappears as an object of attention unless something goes wrong with it. In Polanyi's terms, the child is now attending *through* his body *to* the world beyond. He feels a growing mastery.

Friedrich Nietzsche said that joy is the feeling of one's power increasing. This needn't be understood as the motto of an insatiable tyrant. It captures something important about the role that skill plays in a good life. When we become competent in some skilled action, the very elements of the world that were initially sources of frustration become elements of a self that has expanded, by analogy with the way a toddler expands into his own body and comes to inhabit it comfortably. And this feels good.

RIDING MOTORCYCLES: GYROSCOPIC MAN

We remain capable of learning new skills throughout life. Evolution has endowed us not with a fixed design and scripted behaviors, adapted to cope with a particular environment (securing food and mates in the Pleistocene savannahs, as the evolutionary psychologists keep insisting), but rather with highly plastic neural resources and "an ongoing regime of monitoring and recalibration," as Clark says.[11] When we begin to learn a new skill, new equipment often mediates between our bodies and the world, and the loop of perception through action may take a detour through physical phenomena that are quite alien to the natural human body. In that case, we find ourselves returned to the condition of being like a toddler, figuring out how to maneuver ourselves through the world.

The steering dynamics of a motorcycle are a subtle and astonishing thing. At higher speeds, to make a motorcycle initiate a turn to the left, you apply pressure as though you were trying to turn the handlebars to the right. Motorcyclists call this countersteering, and it is indeed counterintuitive. Turning the handlebars briefly to the right makes the bike lean to the left because of gyroscopic precession, and it is the leaning that accomplishes the turning. (The reader may recall a classroom demonstration of gyroscopic precession involving a spinning bicycle wheel that one holds by its axle, while seated on a chair that is free to swivel. Tilting the axle—that is, trying to rotate it on a horizontal axis perpendicular to the axle itself—is hard to do, and makes the chair you are sitting in swivel on a *vertical* axis. It is a very weird experience; there seem to be demonic forces at work.)

In cornering a motorcycle there is a series of motions and exertions that get installed in muscle memory through practice, and these are integrated with the visual cues of cornering. Once the integration is fairly secure, it is the visual cues that the motorcyclist attends to, not the muscular exertions. The higher the speed, the more intensely they are attended to; the level of concentration involved in motorcycle road racing is truly impressive, as will be appreciated by anyone who has seen a picture of a rider at full lean through a corner. You will see his inside knee, sometimes even inside *elbow*, scraping the ground (racing leathers have plastic pucks on the knees so they don't get torn up). The next time you see such a photo, look for the rider's eyes. If they are visible through the helmet's visor, you will see them directed nearly perpendicular to the bike's direction of travel, as the rider looks all the way through the corner.

This brings up another uncanny fact about motorcycle steering: the bike goes wherever your gaze is focused. Most important, if your eyes lock on some hazard in the road, you will surely hit it. This is not a superstitious motorcyclist's version of Murphy's Law; it is a reliable fact, and it reveals something deep about the "intentionality" of our prereflective sensorimotor negotiation of the world. Inhabiting the kind of bodies that we do, our gaze and our locomotion are connected in ways that work for us, and we don't have to think about it. But this accomplished integration becomes a liability when riding a motorcycle, and must be deliberately short-circuited. You have to learn to unlock your eyes as quickly as possible from every hazard, and instead look where you want to go.

This visual demand is absolutely counterintuitive. When walking, we move away from a hazard (for example a snarling dog) while keeping it in view. Our action programs, visual system, and "affect" (immediate, visceral judgments of good or bad such as happens when we see a spider) are integrated in a way that is adaptive for us, and have achieved a certain automaticity. But when the relation of your body to the world is mediated by a machine, one that requires a very different set of muscle responses to achieve the desired avoidance, then you aren't well adapted until you have reintegrated muscle response, visual system, and affect into a very different collection of automated responses. At the heart of this learning process for the motorcyclist is a phenomenon utterly unknown to the natural human body, namely gyroscopic precession.

Gibson's work sheds light on all this. He suggested that the concept of an "ecological niche" is necessary to properly understand perception. A niche is not quite the same as a habitat. A niche "refers more to *how* an animal lives than to *where* it lives."[12] It is not simply the physical surroundings, but the aspects of those surroundings that are meaningful for an animal given its way of life.[13] When you live on two wheels, gyroscopic precession is as important a feature of your ecological niche as gravity.

Gibson's most interesting and controversial point is that *what* we perceive, in everyday life, is not pure objects of the sort a disinterested observer would perceive, but rather "affordances." The affordances of the environment are "what it *offers* the animal, what it *provides* or *furnishes*, either for good or ill." Affordances elicit and guide action; Gibson suggests they also organize perception. Things in our environment show up in the vivid colors of *good* (for a motorcyclist: a median strip with a curb that is low enough that it could serve as an escape route if things get hairy in front of you) and *bad* (the oily patches that are usually present in the center of a lane at an intersection, where cars drip fluids while idling). As Alva Noë puts it, "When we perceive, we perceive in an idiom of possibilities for movement."[14]

Our perception of these possibilities depends not only on the environmental situation, but also on a person's skill set. A martial artist faced with a belligerent man at a bar sees the way the man is standing, and his distance, as affording certain strikes and foreclosing others, should it become necessary.[15] Because of long practice and habituation, when he looks at the man's stance, this is what he sees. He may also perceive the furniture nearby, and the objects lying within reach on the bar, in terms of their affordances for combat. He sees things that people like you and me don't. That's why we shouldn't mess with him.

Affordances lie in the *fit* between an actor and his or her environment. When that relationship is mediated by a prosthetic, such as a motorcycle, it changes the field of objects that we perceive and how we perceive them. Gibson considered only the natural human body when he investigated the ways movement through our environment influences our perception. But his idea of affordances provides a useful foundation for thinking about culture and technology—that is, for thinking about the distinctly human ecological niches that we create for ourselves. This becomes important for my concept of the situated self, and we will develop it further in the chapter "Encountering Things with Other People."

But for now let us stay focused on motorcycling and try to explicate this one instance of a specialized human ecological niche in fuller detail.

In addition to gyroscopic precession, a further "unnatural" challenge in motorcycling is the categorically different rate at which you are moving toward the things in your visual field, compared with our usual bipedal locomotion. This makes it imperative to keep one's eyes fluid. One has a certain amount of time, typical of the particular environment one inhabits, to judge a hazard and respond accordingly. A ship's captain on the open sea has a lot more time to avoid another ship than an alligator wrestler has to read the signs of his opponent's impending move. At high speeds these "judgments" (if that is the right word) must be very fast. They cannot be the result of a conscious inferential process (just as they are not when we are running to catch a fly ball), because inference is a slow and cognitively costly activity. The subconscious integration of sensorimotor data that one is performing while riding a motorcycle at high speeds requires a great deal of concentration, but not a lot of articulate thinking.

Those data are inextricably bound up with a host of mechanical contingencies. A motorcyclist feels the road through his tires, and racers are very particular about the air pressure, the cross-sectional shape, and the particular rubber compound used. A constant curvature in the tire's cross section will make the bike feel linear in its response to the rider's leanings. A flatter cross section or "profile" will produce a larger contact patch while the motorcyclist is riding upright in a straight line (thus increasing braking power at the end of a straightaway, which is important because then the rider can initiate braking later), while a peakier profile will have a larger contact patch while the cyclist is leaned over in the turns. Choice of tire profiles therefore depends somewhat on the track and a team's race strategy. Different rubber compounds break loose with more or less abruptness, which influences the rate at which the rider can apply throttle coming out of a corner (in this situation traction, not horsepower, is the limiting factor; in general, traction is "conserved" and must be judiciously allocated among lean angle, acceleration, and braking). Some tires communicate better than others. The rider also feels the road through the suspension, which on race bikes is separately adjustable for compression and rebound damping, as well as effective spring rate. Some racers are particular even about the steering head bearings, claiming they can feel the difference between tapered roller bearings and ball bearings. Needless to say, a racer does not attend to any of these elements while leaned over at 130 miles an hour in a corner, knee on the ground, separated by a few feet from other riders in a pack. (Such is the mutual trust of skilled professionals.) The mechanical contingencies of traction and gyroscopic precession become second nature, and are given no more thought than the hockey player gives to his stick while he is playing, or the blind man gives to his cane.

The philosopher Adrian Cussins writes about two different ways of knowing about speed.[16] He relates the experience of riding his motorcycle around London, adjusting his speed in response to various weather and traffic conditions, and contrasts this with the way one knows one's speed when one reads it off a dial or digital readout. In that case, speed is rendered as a number, and to learn the significance of this number one has to compare it with another number: the posted speed limit. But this "significance" is of a much thinner kind. Cussins's point is that knowing one's speed in this second way is to render speed as a proposition: I know *that* I am going forty-five miles per hour. This is a fact—the sort that has objective validity, but is divorced from the particular driving situation in which the motorcyclist finds himself. "Forty-five miles per hour" is not speed, it is a representation of speed. It has the virtue of standardization; it is the kind of fact that is transportable "from one embodied and environmentally specific situation to another."

These two ways of knowing about speed "are taken up in very different, sometimes competing, cognitive orientations to the world," as Cussins puts it. When the objective representation of speed interposes itself between the motorcyclist and his perception of his situation, it can interfere with his direct world-inhabiting. Cussins writes that "the great advantage of experiential content is that its links to action are direct, and do not need to be mediated by time-consuming—and activity-distancing—inferential work."

If Cussins is right, reliance on a speedometer tends to subtly bump us out of a skillful way of driving, and this is due to the interference of objective knowledge with experiential knowledge. But Cussins doesn't elaborate how this interference might happen. Following some clues in the cognitive science literature, I'd like to suggest that the interference is due to a substitution that occurs, wherein the symbolic representation of speed becomes an object of attention, displacing somewhat the ecological, sensorimotor experience of speed. Crucially, unlike the ecological experience, the symbolic representation of speed is "affect-neutral" (it isn't scary), so it doesn't *prime* the action programs (here, evasive maneuvers) that for an experienced rider have become integrated with his threat perception and have achieved a certain level of automaticity.

The negative affordances that a motorcyclist sees aren't limited to things like oily spots on the pavement. The road is, after all, a social place. You do well to notice the brunette in the short skirt standing at the intersection, because the guy driving the car in front of you may slam on the brakes. But the old lady following closely in the car behind you won't. You have been watching the old lady with interest. As far as you can make out in your vibrating mirror, she has a look of sour disapproval on her face, and it is directed at your taillight. You also notice the driver of the delivery truck that has just appeared at a side street. He appears to be laughing: he is engaged in a conversation with someone else in the cab of the truck, seated on the side opposite to the side that you are approaching from. The driver looks like just the slovenly sort who would pull out without double-checking. (Motorcyclists become ethnographers of necessity, or rather rank stereotypers, for the same reason that cops do: they face risk. Stereotyping is efficient for snap judgments.) Having been scared many times in the past, you are attuned to the kind of information that is important when riding in urban areas, and this information is different in kind from your instrument-read speed.

I suspect overreliance on the speedometer may slacken the bonds between action and perception, and indeed there is evidence that when our attention is diverted to symbolic representations, it can have such a loosening effect. This is what appears to be going on with a chimpanzee named Sheba who has been taught numerals. Andy Clark tells her story:

> Sheba sits with Sarah (another chimp), and two plates of treats are shown. What Sheba points to, Sarah gets. Sheba always points to the greater pile, thus getting less. She visibly hates this result but can't seem to improve. However, when the treats arrive in containers with a cover bearing numerals on top, the spell is broken, and Sheba points to the smaller number, thus gaining *more* treats.

The interpretation is that the numerals, because they don't look tasty, "allow the chimps to sidestep the capture of their own behavior by ecologically specific, fast-and-frugal subroutines." They provide "a new target for selective attention and a new fulcrum for the control of action."[17]

Abstracting from the concrete objects of their environment, the chimps become better at maximizing their utility. They become more Protestant, we might say: to get maximum treats in the future requires a bit of asceticism for the moment, and this becomes possible if you redirect your attention to something abstract, such as money (for the Protestant) or numerals (for the chimp). The abstract thing becomes "a new fulcrum for the control of action," as Clark says. Thus is born civilization.

But when it comes to motorcycling, "ecologically specific, fast-and-frugal subroutines" are mostly a good thing, because everything happens very fast.[18] Let us call them perception-action circuits. Just as reaching is triggered by the sight of something tasty for chimps and other non-Protestants, the solicitations of the motorcycling-specific environment trigger steering and other control inputs for the rider. These circuits are tied to *affect*: the kind of response you have to the sight of something tasty or something dangerous. In the case of a learned skill, these perception-action-affect circuits represent an achieved integration, and serve as the foundation for fluid, relatively effortless performance.

Is there a role for explicit thinking in this kind of performance?

THE ROLE OF LANGUAGE IN ACQUIRING SKILL UNDER CONDITIONS OF RISK

Among philosophers there is currently a quarrel about what role (if any) concepts—the kind you can state in language—play in skilled activities.[19] On one side is Hubert Dreyfus, who says that when we are engaged in an activity that we are already competent in, anything so thoroughly mental as a "concept" doesn't normally play a role. It can only get in the way and disrupt our "smooth coping." By "smooth coping" he means a way of acting where our responses to the things we are dealing with are elicited from us by the situation, without articulate thought. This is how you tie your shoes in the morning, for example. It is a skill you learned long ago, and has become automatic. Dreyfus offers this idea as a corrective to the view that our actions are always caused by prior "mental" operations.

On the other side is John McDowell, who offers what I take to be an important countercorrective to the "smooth coping" notion by emphasizing the role of concepts in skilled activity. We don't shut off our thinking, the way Dreyfus seems to suggest. Though McDowell doesn't mention the fact, I believe his emphasis on the role of conceptual thinking is especially necessary when the activity in question is dangerous, and there are contingencies that remain beyond your control no matter how skilled you are, as in motorcycling. (Motorcycling is different from tying your shoes in this regard.) We needn't enter into the minutiae of this academic debate, only know that it hovers in the background of what follows.

Bernt Spiegel is a German automotive psychologist who spent much of his career consulting for Porsche. He was also a motorcycle road race instructor. His book *The Upper Half of the Motorcycle*, an unobtrusive masterpiece, makes no reference to the recent literatures on embodied cognition or "nonconceptual mental content," but is overflowing with observations that contribute to these lines of inquiry. He writes, "One simply has to *know* about some situations before behavior can be adapted on the basis of this knowledge."[20] He gives the example of seeing your own shadow directly in front of the bike on the road. This happens to occur at just those times when you have excellent visibility ahead, and therefore feel especially relaxed. But the sight of your shadow should trigger an alarm: oncoming traffic or motorists waiting to cross in front of you likely cannot see you. It doesn't *feel* dangerous; you have to have a theory about the situation, as it were. That is, to ride safely you have to actively summon a bit of knowledge that can be stated as a proposition, using language. Initially, this knowledge is abstract and obtrusive; it interferes with your flow. But this theoretical knowledge becomes integrated with your perception-action-affect circuits after you have had a number of close calls. (This is my own assertion, not Spiegel's.)

When you nearly get creamed on the road, there is an intense physiological reaction that occurs. I experience it as my stomach leaping up toward my solar plexus, and this happens immediately, even as the event is occurring. It is followed soon afterward by a shaking that is presumably caused by adrenaline, and extreme weakness in my throttle hand, which is sometimes debilitating. This has a powerful effect: an association between the pertinent features of the situation you just lived through and DANGER gets burned into the circuits of embodied cognition that you rely on while riding. Suppose one feature of this situation is the sight of your own shadow in front of you. After a close call, the association between that sight and bodily danger becomes not a mere proposition, but something you really know and feel. No inference is necessary.

But I suspect this integration occurs only if you have the proposition in hand ahead of time, so that you have identified the sight of your own shadow in front of you as an element of "the situation." There are an indefinite number of true facts that could be stated about the circumstances leading up to your close call: the color of the car that almost hits you, the phase of the moon, and so on ad infinitum. The role of attention in the exercise of a skill is to pick out those features of the scene that are pragmatically significant and that therefore, taken together, define "the situation." In the early stages of learning a skill, explicit propositional knowledge, stated in language for instructional purposes, plays a crucial role in directing attention toward its proper objects. Bathed with attention in this way, these objects then become *available* for integration with affect and action routines by the subconscious mind, following a close call.

Note that the role played by language implies that achieving competence—even in an activity as solitary as motorcycling—has an important social dimension. You learn things from others by reading books, having conversations with other practitioners, or watching tutorials on YouTube. Gary Klein has famously studied the decision making of firefighters, and discusses their ability to discern when a building is about to collapse, allowing them to get out in the nick of time. Klein emphasizes their ability to integrate subtle sensual data and recognize patterns. But as far as I know he does not address the role of affect, nor the role that language plays (on my account) in priming the integration of affect with perception and action. Presumably, the more experienced firefighters try, at least, to describe to a beginner the peculiar sequence and combination of sights, sounds, and feelings that precede a building collapse, so that the beginner knows what to be alert to in the chaos of a burning building. If my hypothesis is correct, such explicit instruction would be an important preparation for the establishment of those perception-action-affect circuits which, once integrated, become the basis for high-level performance.

The cohesiveness and ongoing association of a firefighting unit offer an advantage not enjoyed by most motorcyclists: they are under mutual surveillance and can criticize one another's mistakes. They can also cover one another's blind spots, offering up a third-person perspective such as "There was a large ember floating upward right behind you as you exited that room. You got lucky." Such facts, conveyed by a colleague, can become material for a firefighter's retrospective understanding of "the situation," or indeed a collaborative reconstruction of it. His own experience is altered in conversation.

That is, the array of sensual data that count as pragmatically relevant for grasping the situation may be expanded or shifted through a kind of triangulation with others who were not merely there, but engaged in the same task, facing the same dangers. You debrief one another. The fruit of this conversation enters into your ongoing rehearsal of the experience. If this rehearsed version bears up, and jibes with further experience, it becomes internalized, available to the subconscious mind in coping with future situations. For experiences to become part of the secure, sedimented foundation of a skill, they must be *criticized*. Other people (and the resources of language) are indispensable. Without them, your experiences are partial, and may sediment as idiosyncratic bad habits.

The power of these conversations to clarify your experience, rather than introduce fresh confusions, depends in part on the dialectical abilities of your colleagues. They have to be able to interrogate their own experience of the fire critically, and bring their experience into the conversation in such a way that their initial interpretation of it is put at risk. They have to be capable of offering it up, without undue attachment, to the shared enterprise of trying to understand structure fires. In other words, they must have the art of philosophical conversation (which is a kind of moral accomplishment). I believe the most competent people in any field do have this art to some degree, though they probably wouldn't name it as such.

Getting things right requires triangulating with other people. Psychologists therefore would do well to ask whether "metacognition" (thinking critically about your own thinking) is at bottom a social phenomenon. It typically happens in conversation—not idle chitchat, but the kind that aims to get to the bottom of things. I call this an "art" because it requires both tact and doggedness. And I call it a moral accomplishment because to be good at this kind of conversation you have to love the truth more than you love your own current state of understanding. This is, of course, an unusual priority to have, which may help to account for the rarity of real mastery in any pursuit.

Another reason to remain self-critical while motorcycling is that sometimes there are features of a situation that present no sensual cues at all, for example the patch of gravel on the road around the blind curve you are just now leaning into. When you are dealing with potential hazards that aren't present to perception, you have to actively form *hypotheses* about bad contingencies, and project them out into the world. Doing so primes the appropriate action plans, and makes them more quickly accessible. Imagining what *could* happen is an important role for the conscious mind, so it must stay involved. Being in a state of "flow" without such worries sometimes makes you feel like Superman, but it is easy to flow yourself right into the truck that has drifted into your lane around the blind curve ahead.

Isn't it true in general that life is shot through with hidden contingencies? Risk is present in any activity that is directed to some goal—the risk of failure. Unlike animals that live in the moment, guided by instinct, we are constantly monitoring our own performance, second-guessing it, tuning it up.

Spiegel points out that when riding with limited visibility (which is usually the case), we tend to ride according to a "risk composite." We are vaguely aware of some really bad contingencies, but we also know that the odds of these actually occurring are small. Our response, as folk statisticians, is to slow down *a little bit*, as though we are weighting the bad contingency by multiplying it with its unlikelihood, and thereby arriving at the appropriate speed. But this is self-delusion. Our reduced speed is nowhere near as slow as it would need to be to avoid the hazard if it should occur. (Then again, without self-delusion we'd probably never enjoy ourselves.)

Riding in the controlled environment of a racetrack is fundamentally different from riding on the street, in terms of the mental practice it involves. I tagged along with the MotoGP tour through its first two races of the 2013 season (in Doha, Qatar, and Austin, Texas) and talked to some of the riders. On the track, to be competitive one has to have complete faith in the mental image one has of a corner. If there is a rise in the track, blocking your view through the turn (for example Turn One at Austin's Circuit of the Americas), this mental image of the corner will extend beyond what is *perceptually* present when you initiate the turn and commit to a certain line. Your mental image is based on repetition (you go around the same track many times in practice sessions) and is assumed to be reliable. In the event of a disrupting hazard, there are corner workers who wave yellow flags and position themselves where you can see them early. These corner workers serve to relieve the conscious mind of its burden of actively positing hypotheses about bad contingencies. A masterful road racer, thus relieved, takes the art of motorcycling to its highest level. It is beautiful to watch, and forces one to recalibrate one's sense of what human beings are capable of.

On the street, riding like Marc Márquez (the current MotoGP champion) is an ideal that is in fruitful tension with the demand for continued vigilance by the conscious mind. Managing this tension is itself an art. As Spiegel puts it, the role for the conscious mind is "alert watchfulness, without meddling." It is "an unstable condition, which degrades all too easily into either a complete lack of watchfulness or too much involvement." When this mental practice is lived, it doesn't manifest as something beautiful for a bystander to behold, as road racing does. Rather, it becomes impressive only as a cumulative accomplishment in the life of the rider, measured in miles ridden without incident but including also some all-too-brief stretches of Márquez-esque transcendence.

This "alert watchfulness without meddling" by the conscious mind while one is riding on the street often takes the form of hunches: hypotheses about what might happen that are conscious but not fully articulate, because they don't need to be. You recognize a familiar situation: there are strip malls on either side of a major thoroughfare, each with entries to the main road. The street numbers are posted only erratically, on haphazard buildings set far back from the main road. The car in front of you slows down, then speeds up, repeatedly. Hypothesis: this person is looking for a particular business, and when he spots it he may quickly veer across two lanes to get to it. Your motor responses are cocked and loaded, as it were, because you recognize the pattern.

Recall the old story about the prisoners who have been together so long they know all the jokes in circulation and have numbered them. Eventually, the telling of a particular joke consists of simply yelling out "seven," at which point everybody cracks up.

Similarly, when an experienced motorcyclist rides through a zone of strip malls, the descriptive complexity of the scene has been reduced to a type for him. When I first started riding, I remember being tense in such situations, besieged by unpredictability and an overwhelming variety of data that needed to be monitored. At some point my vigilance became more relaxed, more enjoyable, and also more effective. I noticed that I was riding with my left thumb poised on the horn button. My right hand is indexed to the front brake lever by the two fingers that rest lightly on it. My upper body is soft, primed for quick steering inputs. This riding posture *is* a kind of hypothesis, a provisional understanding of what will occur, installed in my body and ready to be deployed. Once this hypothesis/posture is set, it seems to free me up to ride loosely, without being overly taxed cognitively.

I have a little verbal formula that I say out loud when I enter such riding situations: "They want to kill you."[21] (Wearing a full-face helmet, you can hear your own voice booming over the road noise.) Similarly, road racers sometimes put stickers with verbal phrases on their gas tanks, such as "Look deep into corners" and "Use a late apex." Such mottos are taught to the beginner, and for him they have the status of precepts. Why would an expert continue to use them? Andy Clark writes that even for experts, "verbal rehearsal supports a kind of perceptual restructuring via the controlled disposition of attention . . ."[22] We use these verbal prompts to maintain performance in challenging circumstances.

Sometimes a verbal expression will serve as a "tag" for a particular feeling that an expert has become attuned to, an elusive state that he wishes to achieve. In these cases, the phrase will be mysterious when used as an instructional device for the novice. Spiegel gives the example of feeling your consciousness "flowing down through the contact patches." (Once again, these are the patches of rubber where the tires make contact with the road; they change shape with different lean angles.) The formula may sound like mysticism, but is meant to capture what it feels like when your riding is totally dialed in and the bike has been incorporated into your extended body. Similarly, a pianist reports initially being frustrated by his teacher's instructions to "get the time into the fingers" or use "jazz hands." At first this is unhelpful, but eventually, "what seemed like just vague words to the novice has now become very detailed practical talk, a shorthand compendium of 'caretaking practices' for toning and reshaping the grooved routines."[23]

This drive to continually tone and shape up a skill is lost sight of if we take tying one's shoes as the paradigm of skilled action. That is an activity for which we adopt a "sufficing" standard: Is the shoe tied or not? Being able to tie your shoe is a secure accomplishment, a state of stasis. But in activities that we take seriously, such as music and sports and going fast, we strive for excellence. Unlike animals that live in the moment and merely cope with their world (however smoothly), we are erotic: we are drawn out of our present selves toward some *more* skilled future self that we emulate. What it means to be erotic is that we are never fully at home in the world. We are always "on our way." Or perhaps we should say that this state of being on our way to somewhere else is our peculiar human way of being here in the world.

We have considered how perception is intimately bound up with action, and what it is like to attend to the world through implements that get incorporated into our consciousness. The boundaries of the self seem to expand. As we push out into the world, first as toddlers and then as tool users, we perceive it differently because we are now inhabiting it in a more determinate way, conditioned by the particulars of the skill and the implements we use. Through a skilled practice, the self has been brought into a relation of fit to the world. And this can be quite absorbing.

To emphasize the role that our bodies play in determining how we inhabit and therefore perceive the world, and to entertain the notion of cognitive extension, is to put oneself on a collision course with the central tenets of the official anthropology of the West. As we have already noted, embodied perception poses a direct challenge to the idea that representation is the fundamental mental process by which we apprehend the world. But these developments in psychology also pose a challenge to contemporary ethics. And this is no mere coincidence. In ethics, as in epistemology, the idea of representation is central to our Enlightenment worldview. Immanuel Kant insisted that in order to avoid special pleading, in moral reasoning we should regard ourselves and others not as individuals but as representatives of the genus "rational being," and approach one another through the filter of this abstraction. In Part II, "Other People," we will consider the flavor that this abstraction imparts to our interactions with one another, and contrast it with the kind of interaction that is made possible by an "ethics of attention" in which we are alive to the concrete particularity of others.

We have more work to do to establish what it means to encounter *things*, and the next few chapters will remain focused on this. But I want to start building the bridge that will get us from things to people, and from epistemology to ethics. This bridge begins to take shape when we notice that the technological concept of virtual reality also expresses a moral ideal. More precisely, it expresses what philosophers call a metaethical position. That is, it carries with it a certain understanding of the underpinnings of ethics: a picture of the moral agent and how she stands in relation to the world beyond her head. Taken in this sense, virtual reality provides an especially clear point of contrast to the concept of the situated self that I am offering.

3

VIRTUAL REALITY AS MORAL IDEAL

If you have children, you know that the will of a toddler has a kind of purity to it: he wants what he wants, and refuses what he refuses, without reference to any fact that might inconvenience his will. It is freezing outside, but he doesn't want to wear shoes to the park. By contrast, the will of an adult is shaped by his interactions with material reality. To say that the will is shaped by the world means that putting on shoes, for example, is no longer experienced as an accommodation. It is just what one does, as a condition for the possibility of doing anything else.

Beyond such minimal conditions, we enter the realm of skill. In skilled action, there are a lot of contingent facts about *stuff* that have to be learned at a deep level if we are to achieve our purposes: catching a fly ball, hitting a slap shot, cornering a motorcycle. Further, these are purposes we didn't even have before we began our initiation into the skill, and started to perceive the affordances that this skill revealed in our environment. Acquiring skills, we acquire new motivations; a new space of reasons for action.

The adult will is not something self-contained; it is situated in, and formed by, the contingencies of the world beyond one's head. The kind of self that accepts this elemental fact contrasts with, and therefore brings into clarifying relief, the more fragile kind of self that is posited in contemporary ethics and fostered by contemporary technology. The freedom and dignity of this modern self depend on its being insulated from contingency—by layers of representation.

As Thomas de Zengotita points out in his beautiful book *Mediated*, representations are *addressed* to us, unlike dumb nature, which just sits there. They are fundamentally flattering, placing each of us at the center of a little "me-world."[1] If the world encountered as something *distinct* from the self plays a crucial role for a person in achieving adult agency, then it figures that when our encounters with the world are increasingly mediated by representations that soften this boundary, this will have some effect on the kind of selves we become. To see this, consider children's television.

THE MOUSEKE-DOER

In the old Mickey Mouse cartoons from the early and middle decades of the twentieth century, by far the most prominent source of hilarity is the capacity of material stuff to generate frustration, or rather demonic violence. Fold-down beds, ironing boards, waves at the beach, trailers (especially when Goofy is at the wheel of the towing vehicle, on a twisty mountain road), anything electric, anything elastic, anything that can become a projectile. Anything that can suffer termite damage that remains hidden until the crucial moment. Springs are especially treacherous, as are retractable blinds. Snowballs can be counted on to grow by a couple of orders of magnitude on their way down the slope toward your head. At any given moment, the odds of being seized by the collar by a severely overwound grandfather clock are nontrivial. Icicles: don't stand anywhere near them. Bicycles tend to become unicycles, unpredictably, and rubber cement is easily mistaken for baking powder. Why do they have nearly identical labels?

These early cartoons present a rich phenomenology of what it is like to be an embodied agent in a world of artifacts and inexorable physical laws. The tendency of these things to thwart the human will is exaggerated, and through exaggeration a certain truth gets brought forward. As the stand-up comics say, only the truth is funny. In depicting the heteronomy that the world of objects inflicts on us, the slapstick sufferings of Donald Duck acknowledge, and thereby seem to affirm, the human condition as it is, beneath the various idealisms that would transport us out of that condition.

The Disney cartoon franchise now has many departments. One of them, *Mickey Mouse Clubhouse*, on the Disney Junior network, retains the same characters. But the difference in how material reality is presented could not be more stark, and in this difference a shift in the relation between self and world becomes evident as well.

Each episode begins with Mickey looking into the camera and speaking directly to the viewer in tones of solicitous hypercongeniality, pausing every so often to elicit a response as he cups his hand to his ear. Once he has performed this ritual of seeking consent, he speaks some magic words ("Miska, Mooska . . .") and the Clubhouse rises up out of the ground in a psychedelically abstract, parklike landscape. There, the characters present themselves for review, each auditioning for the viewer's consideration with his or her own brand of delightfulness.

The Clubhouse is filled with amazing technology that always works perfectly. In the episode "Minnie's Mouseke-Calendar," a strong wind is blowing. You might think this is the setup for some slapstick. But when Goofy starts to get blown away, a retractable hand rises up out of a trapdoor (disguised as a paving stone) and gently pulls him back down to the earth.

The current episodes are all oriented not around frustration but around solving a problem. One does this by saying, "Oh Toodles!" This makes the Handy Dandy machine appear, a computer-like thing that condenses out of the Cloud and presents a menu of four "Mouseke-tools" on a screen, by the use of which the viewer is encouraged to be a "Mouseke-doer."

In the episode "Little Parade" some wind-up toy marching band figures have been overwound and scattered, and must be retrieved. One of them ended up on the other side of a river that runs beneath the cliff Goofy is standing on. Goofy says the magic words. The Handy Dandy machine boots up and presents its menu of options; one of them is a giant slide. Perfect! The slide is conjured out of the ether and settles gently into place to run from the cliff to the far bank, where Goofy retrieves the errant toy.

There are four problems per episode, and each can be solved using one of the four tools. This assurance is baked into the initial setup of the episode; no moment of helplessness is allowed to arise. There is never an insoluble problem, that is, a deep conflict between the will and the world. I suspect that is one reason these episodes are not just unfunny, but somehow the opposite of funny. Like most children's television these days, *Mickey Mouse Clubhouse* is doggedly devoted not to capturing experience, that is, to psychological truth, but to psychological adjustment. It is not a depiction so much as an intervention—on behalf of parents, teachers, and others who must manage children.[2] The well-adjusted child doesn't give in to frustration; he asks for help ("Oh Tootles!") and avails himself of the ready-made solutions that are presented to him.

To be a Mouseke-doer is to abstract from material reality as depicted in those early Disney cartoons, where we see the flip side of affordances. Perhaps we should call unwanted projectiles, demonic springs, and all such hazards "negative affordances." The thing is, you can't have the positive without the negative; they are two sides of the same coin. The world in which we acquire skill as embodied agents is precisely that world in which we are subject to the heteronomy of things; the hazards of material reality. To pursue the fantasy of escaping heteronomy through abstraction is to give up on skill, and therefore to substitute technology-as-magic for the possibility of real agency.

This cartoon magic may be fanciful, but one would be hard-pressed to find any meaningful distinction between it and the utopian vision by which Silicon Valley is actively reshaping our world. As we "build a smarter planet" (as the IBM advertisements say), the world will become as frictionless as thought itself; "smartness" will subdue dumb nature. But perhaps even thinking will become unnecessary: a fully smart technology should be able to *leap in* and anticipate our will, using algorithms that discover the person revealed by our previous behavior. The hope seems to be that we will incorporate a Handy Dandy machine into our psyches at a basic level, perhaps through some kind of wearable or implantable device, so that the world will adjust itself to our needs automatically and the discomfiting awareness of objects as being independent of the self will never be allowed to arise in the first place.

The appeal of magic is that it promises to render objects plastic to the will without one's getting too entangled with them. Treated from arm's length, the object can issue no challenge to the self. According to Freud, this is precisely the condition of the narcissist: he treats objects as props for his fragile ego and has an uncertain grasp of them as having a reality of their own. The clearest contrast to the narcissist that I can think of is the repairman, who must subordinate himself to the broken washing machine, listen to it with patience, notice its symptoms, and then act accordingly. He cannot treat it abstractly; the kind of agency he exhibits is not at all magical.

The creeping substitution of virtual reality for reality is a prominent feature of contemporary life, but it also has deep antecedents in Western thought. It is a cultural project that is unfolding along lines that Immanuel Kant sketched for us: trying to establish the autonomy of the will by filtering material reality through abstractions.

KANT'S METAPHYSICS OF FREEDOM

"Autonomy of the will is the property of the will through which it is a law to itself *independently of all properties of the objects of volition*," Kant writes. "If the will seeks that which should determine it . . . in the constitution of any of its objects, then heteronomy always comes out of this." In such a case "the will does not give itself the law but the object through its relation to the will gives the law to it." Autonomy requires that we "abstract from all objects to this extent—they should be without any influence at all on the will so that [the will] may not merely administer an alien interest but may simply manifest its own sovereign authority as the supreme maker of the law."[3]

These pronouncements only make sense if we recover some

historical context and understand them as a response to a problem. Consider the alarm that must have come naturally to thoughtful people beginning in the seventeenth century. The new natural science offered a mechanistic account of nature from which there seemed no reason to exempt human beings.[4] The naturalistic psychology of Thomas Hobbes and others threatened to subsume human freedom to the deterministic realm of material causation. Thus "free will" became a problem that had to be addressed. The foundations of morality seemed to be at stake. In his *Groundwork of the Metaphysic of Morals*, Kant means to put the freedom of the will on a new footing, where it will float free of all natural necessities.

As the title of the *Groundwork* suggests, Kant tries to lay out not the actual content of moral principles, much less a detailed account of our obligations to others, but rather what *kind* of thing morality is. He insists that it exists in the realm of the ideal, not the empirical. This is crucial for the possibility of moral freedom, as against naturalistic determinism. So it is not a realistic picture of experience so much as a lawyerly argument, in which Kant tries to construct a fortress in which moral freedom is not threatened by the dumb causation of Newtonian nature. But this leads Kant into some strange assertions.

"The object *through its relation to the will*" cannot be allowed to determine the will, if it is to be free. I take Kant to be talking about relations of *fit* between a person and the field of objects that he deals with in his environment—precisely what we explored under the heading of "affordances" and "ecological niches" in the previous chapter. Kant builds a high wall between the empirical world and the purely intellectual, where we discover a priori moral laws. Reasons to act must come only from the latter if we are to be free, and the will is to remain pure, "unconditioned" by anything external to it.

But in our discussion of affordances and cognitive extension, we learned that we may acquire purposes *through* initiation into a skill, such that affordances perceived in the environment provide not just handles for actions previously decided upon by pure reason, but new motivations; a new space of reasons for action. This is precisely what Kant calls heteronomy, if I have understood him correctly, and I call the foundation of human agency as we actually experience it.

The challenge posed by cognitive extension to contemporary culture cuts deep, because Kant's metaphysics of freedom is at the very core of our modern understanding of how we relate to the world beyond our heads. *Mickey Mouse Clubhouse* merely presents an exaggerated version of this understanding, and like most caricatures it brings to prominence the defining features of the thing caricatured.

The Handy Dandy machine would teach us that it is not essential to human agency that you understand *how* your choice is realized or that you in fact *do* anything to realize it. The intervention of magic allows the moment of choice to be isolated from the (mysterious) process by which that choice is to be made effective in the world. It is also isolated from what comes before choice; the options for action that you are to choose from are already there, presented to you without involvement on your part. In the previous chapter we learned that, on the contrary, your skill set determines what possibilities for action you perceive in your environment, as in the case of the trained martial artist. How we act is not determined in an isolated moment of choice; it is powerfully ordered by how we perceive the situation, how we are attuned to it, and this is very much a function of our previous history of shaping ourselves to the world in a particular way.

When Kant says the following, it seems to be his take on the martial artist who perceives affordances for combat in the stance of a belligerent man at the bar: "The will does not give itself the law, but an alien impulsion does so through the medium of the subject's own nature as tuned for its reception." The alien impulsion here is the motivational force that some concrete situation has for someone who is "tuned for its reception." Such attunement is heteronomy, for Kant.

To be sure, this way of putting the matter seems apt if you focus on the case of a martial artist who responds to every solicitation or affordance for combat that arises in his environment and goes around karate-chopping everything and everybody in sight. Such a person would indeed be blameworthy in a way that Kant's account of heteronomy captures well; he would be a kind of automaton. But it seems an extreme reaction to take the possibility of such a case as compelling us to ignore the mutual entanglement of will and world, or to posit their segregation as an ideal.

Kant does concern himself with the kind of sensitivity to the world that requires experience, but he treats this in a separate book, the *Critique of Judgment*, and it is significant that he segregates this topic from his argument about the metaphysics of the will's freedom. Moments of attending to the world are separable from the moment of moral choice, and indeed the freedom of the will depends on such separation. Experience is always contingent and particular, and for that reason "unfitted to serve as a ground of moral laws. The universality with which these laws should hold for all rational beings without exception . . . falls away if their basis is taken from the special constitution of human nature or from the accidental circumstances in which it is placed." To be rational is, for Kant, precisely *not* to be situated in the world.[5]

THE PLIABLE CHOOSER

Whether you regard it as infantile or as the highest achievement of the European mind, what we find in Kant are the philosophical roots of our modern identification of freedom with choice, where choice is understood as a pure flashing forth of the unconditioned will. This is important for understanding our culture because thus understood, choice serves as the central totem of consumer capitalism, and those who present choices to us appear as handmaidens to our own freedom.

When the choosing will is hermetically sealed off from the fuzzy, hard-to-master contingencies of the empirical world, it becomes more "free" in a sense: free for the kind of neurotic dissociation from reality that opens the door wide for others to leap in on our behalf, and present options that are available to us without the world-disclosing effort of skillful engagement. For the Mousekedoer, choosing (from a menu of ready-made solutions) replaces doing, and it follows that such a person should be more pliable to the choice architectures presented to us in mass culture.

The absence of the real from *Mickey Mouse Clubhouse*—indeed, the dissociative or abstract quality of children's television in general these days—makes it an ideal vehicle for psychological adjustment; for constructing and managing the kind of selves that society requires, without meddling interference from the nature of things. The particular adjustments to be carried out will have to be determined by a Disney script supervisor, or some other functionary of the modern self.

The basic thrust of these interventions is not something that Kant *caused*. But when dumb nature is understood to be threatening to our freedom as rational beings, it becomes attractive to construct a virtual reality that will be less so, a benignly *nice* Mickey Mouse Clubhouse where there is no conflict between self and world; no contingency that hasn't been anticipated by the Handy Dandy machine. Kant tries to put the freedom of the will on a footing that secures it against outside influence—so it will be "unconditioned," a law unto itself—but he can do this only by removing the will to a separate realm, from which it can have no causal effect in *this* world, the one governed by Newtonian causation. The fantasy of autonomy comes at the price of impotence.[6]

With this comes fragility—that of a self that can't tolerate conflict and frustration. And this fragility, in turn, makes us more pliable to whoever can present the most enthralling *representations* that save us from a direct confrontation with the world. Being addressed to us, these representations allow us to remain comfortable in a little "me-world" of manufactured experience. If these representations make use of hyperpalatable mental stimuli, the world of regular old experience may come to seem not only frustrating but unbearably drab by comparison.

But notice that somewhere in the vicinity of the ideas we have explored about cognitive extension and embodied agency, there may be a route out of this dependence on manufactured experience. In the case of the short-order cook, the hockey player, and the motorcycle racer, one takes one's bearings from a field of objects external to the self, brings one's actions into conformity with them, and something contingent results—some mix of joy and frustration according to one's skill level and a lot of stuff that is beyond one's control.

These experiences hint at the cultural possibilities of engineering and design. The design of things can facilitate embodied agency or diminish it in ways that lead us further into passivity and dependence.

4

ATTENTION AND DESIGN

John Muir published *How to Keep Your Volkswagen Alive* in 1969. In one of the many asides that enliven the book and give it a countercultural feel, he wonders about the effects of some of the newfangled safety equipment, like seat belts. He writes, "If we all constantly drive as if we are strapped to the front of the car like Aztec sacrifices so we'd be the first thing hit, there would be a lot less accidents."

Attention Assist, the latest electronic offering from Mercedes, is a sign of how far we have come from the driving culture of 1969. It is an option package that includes Brake Assist. If the car in front of you suddenly slows down, the Mercedes brakes for you. This frees you up to be somewhere else, mentally—looking at the navigation screen, maybe, or talking to your hedge fund manager. A television advertisement for Attention Assist features a chinless man with babylike cheeks and a bewildered look on his face saying, "I never saw the truck." (The choice of "truck" over, say, "tricycle" seems wise on the part of the copywriters.) The package also includes Blind Spot Assist, so you no longer have to bother with a head check before drifting into the next lane. The basic design intention guiding Mercedes in the last ten years seems to be that its cars should offer psychic blow jobs to the affluent. Just sit back, relax, and think of something pleasing. The eyes take on a faraway glaze. As for other drivers, there is a certain . . . lack of mutuality.

More broadly, the design of automobiles has tended toward insulation, offering an ever less involving driving experience. The animating ideal seems to be that the driver should be a disembodied observer, moving through a world of objects that present themselves as though on a screen. We have throttle by wire, brake by wire, and electrical assist (versus hydraulic assist) brakes, as well as traction control and antilock brakes that modulate our driving inputs for us. What all this idiot-proofing and abstraction amounts to is a genuine poverty of information reaching the driver. What's more, the information that does get through is presented in a highly mediated way, conveyed by potentiometers and silky smooth servos rather than by the seat of your pants. It is therefore highly discrete, and does not reflect fuzzy, subtle variations. Nor is it sensitive to changes that haven't been anticipated and coded for ahead of time, for example the vibration that might arise from a brake caliper bracket that has come loose or cracked. Perhaps most troubling, the electronic mode of presentation means that information about the state of the car and of the road is competing with information from other electronic devices that may be a lot more interesting.

This fetish of automaticity and disconnection can't be called a tendency of "technology," if we insist that the proper standards of technology are simply those of function. Rather, it is the tendency of a peculiar consumer ethic that has embraced Kant's metaphysics of freedom. Disconnection—pressing a button to make something happen—facilitates an experience of one's own will as something unconditioned by all those contingencies that intervene between an intention and its realization.

The wealth of information presented by an older, harder-edged, and lighter car elicits involvement; you have the palpable sense that it is your *ass* that is going sixty miles an hour. Such existential involvement demands and energizes attention. This is why driving a light, primitive sports car is so exhilarating. In a variation on the old funk dictum, we might say, "Involve your ass, your mind will follow." And conversely, "Free your ass, your mind will wander." I suspect John Muir is right with his image of the Aztec hood ornament: having some skin in the game would seem to be an important safety variable.

As traffic engineers have discovered, our approach to driving is influenced quite a bit by the features of a road. Eric Dumbaugh, a civil and environmental engineer at Texas A&M University, says, "We assume that safety is the result of 'forgiving' roads. We figure straightening out streets and widening shoulders makes a road safe."[1] This turns out to be wrong. When roads look dangerous, people slow down and become more heedful. Consistent with this, the failure to recognize risks and appreciate them is found to contribute more than does divided attention to crashes among novice drivers.[2] But there is a relation between these two: perceived risk increases conscious effort and focuses attention.[3] As with cars, so with roads: the always-near possibility of death by blunt trauma should not be made artificially remote from our consciousness.

Emily Anthes writes that among traffic engineers, "in the last decade or so, a few iconoclasts have begun making roads more hazardous—narrowing them, reducing visibility, and removing curbs, center lines, guardrails, and even traffic signs and signals. These roads, research shows, are home to significantly fewer crashes and traffic fatalities."[4] Reporting the findings of Dumbaugh and of Ian Lockwood, a traffic engineer in Orlando, Anthes writes that having on-street parking or bike lanes makes drivers more careful, as does having buildings that come right up to the street, as this seems to give drivers the sense that others are watching them. It is to be hoped that such a face-to-face environment will pull even the Mercedes driver out of the goings-on in his electronic cockpit.

The design of these shared spaces not only influences public safety, but would also seem to play a more far-reaching role in society, through the kind of moral psychology that they promote. Roads are tacitly pedagogical, as are cars. They can foster circumspection—literally, looking around for others and regarding oneself as an object for others in turn—or a collection of atomized me-worlds. In the latter case, we tend not to encounter others unless we literally collide with them.

THE WORLD IS ITS OWN BEST MODEL

The ESPN article I cited earlier quotes one hockey player as saying, "If I pause to *interpret* what I'm sensing when the puck is on my stick, that extra split second can be the difference between a shot and a goal, a win or a loss or getting my head taken off. So the stick has to feel like a piece of you."

A car that interposes layers of electronic mediation between the driver and the road demands an effort of interpretation by the driver, because each of those layers is based on a representation that has no inherent, necessary relationship to the states being represented. Some committee of engineers had to make a whole series of decisions about how the pedal pressure felt by a driver in a car with brake-by-wire, for example, should map onto the braking force delivered and, crucially, the readiness of the system to keep delivering it. Should the pedal effort change with sustained or heavy braking, to convey the fact that those little DC motors doing the work are getting hot? Brake rotors get hot under heavy use and, in doing so, become less effective. This fact gets conveyed to the driver *in a necessary and lawlike way* with the familiar "brake fade" in conventional hydraulic brakes. What was so deeply disturbing about the Toyota recall episode of 2008, I believe, was the revelation that there was software—convention, language, representation—involved in the brakes. This design problem of disconnection or arbitrariness mirrors a fundamental problem in cognitive science: the symbol-grounding problem.

In the computational theory of mind that prevails in conventional cognitive science, we are assumed to have internal representations of the world, and these representations are built on symbols that are meaningless in themselves; they "encode" features of the world in the same way a computer represents states of affairs with a string of zeroes and ones. The symbol-grounding problem is this: How can arbitrary symbols *take on* meaning? How do they acquire propositional content and reference, such that they say something about the world? The same question is posed in philosophy of language, since after all there is no necessary connection between the sounds we make and what those sounds mean. One can refer the words to a lexicon, but the words used in the lexicon face the same problem; there seems to be an infinite regress of the grounding problem.

Embodied representations, as opposed to symbolic representations, do not face this problem. This has implications for the design of automobiles and any other instrument-implement that we use both to perceive and to act on the world. As Arthur Glenberg writes in the article I cited earlier, "embodied representations do not need to be mapped onto the world to become meaningful because they arise from the world." They are "directly grounded by virtue of being lawfully and analogically related to properties of the world and how those properties are transduced by perceptual-action systems."[5] To invoke once more the motto of the new wave of robotics: The world is its own best model.

A harder-edged car, without electronics mediating between action and perception, and in which mechanical noises are not fully damped out, preserves "cross-modal binding," thought by some to be the key to our grasp of reality. Information that we pick up through different senses gets bound together, and coheres in our apprehension of some state of affairs in the world, *because* these various information streams are locked into a common experience of *time*. That is, they co-occur.[6] The rhythmic signature of a brake rotor that has become warped due to overheating is *felt* as undulating pedal pressure, the frequency of which varies with speed. The same oscillation usually shows up as a faint *sound*: that of the rotor moving back and forth, out of the plane of its intended rotation within the brake caliper. Rather than working smoothly together, first one brake pad and then the other gets the brunt of the force. Under heavy braking (for example while descending a mountain road with a trailer), you will also catch a whiff of that unmistakable *smell* of burning brake pad lining. This smell has a different rhythm than the pedal pulse and the sound: its intensity builds slowly, and conveys something like "Things aren't getting any better down here. Thought you might want to know." This too is part of the time-locked stream of information, with varying time signatures, that makes our brains "bind" our various senses together and decide that this is not a dream or hallucination. There is indeed a "thing in itself" out there beyond our heads, revealed by coherent sensory patterns. But only if those patterns are preserved and conveyed to us.

Infants, and robots that take the learning process of infants as their inspiration, generate time-locked patterns of sensorimotor stimulation for themselves by poking at things, manipulating them, and so on. The *sight* of your own hand moving through space gets bound to the *feeling* of this action. The child develops an embodied self-awareness through learning the properties of objects. Different objects resist his body in various ways (light or heavy, soft or hard, slippery or sticky, and so forth), yielding different time-locked bundlings of sensorimotor experience, corresponding to different classes of interaction. This has been shown to play a role in the infant's learning of categories and formation of concepts. Commenting on this literature, Andy Clark writes that "the key to such developing capabilities is the robot's or infant's capacity to maintain coordinated sensorimotor engagement with its environment."[7] A driving experience that provides impoverished feedback limits such engagement, and would seem to promote a kind of regression—back into the womb. Let me concede that this can be nice, especially on a long drive on the interstate. The ideal thing would be to enter a coma. Or perhaps to be like the passive kitten on the carousel.

But only if the interstate is straight. I once had the scary experience of driving a borrowed Toyota Avalon down from the Colorado Rockies on curvy Interstate 70. The Avalon is Toyota's luxury cruiser. I felt so divorced from the road, and from the car, that I found I had to engage in some unaccustomed cognitive work just to keep the thing pointed in the right direction. Traffic was heavy but also fast, moving at 70–75 miles per hour, and keeping up with the flow was an exhausting, white-knuckle experience. I felt like I was guessing the whole time, and even after thirty miles was still constantly surprised by the results of my steering inputs. I began to feel more sympathy for the slow-driving seniors who typically drive these things. My experience corresponds pretty well to this description of using a virtual reality system circa 2001: users peer out at the world that is presented,

> figure out what's going on, decide on some course of action, and enact it through the narrow interface of the keyboard or the data-glove, carefully monitoring the result to see if it turns out the way they expected. Our experience of the everyday world is not of that sort. There is no homunculus sitting inside our heads, staring out at the world through our eyes, enacting some plan of action by manipulating our hands and checking carefully to make sure we don't overshoot when reaching for the coffee cup.[8]

Our embodied mode of existence has given rise to exquisitely sensitive capacities for detecting and negotiating the world, and a good design principle would be to try to exploit these capacities, rather than to sever the connections between perception and action, as the current generation of automotive engineers seems intent on doing.[9]

Mercedes recently unveiled a prototype "enhanced reality" windshield that overlays a digital version of your environment in front of you. BMW, a company that until recently was exemplary in preserving the bonds between car and driver, now gives us fake engine sounds, piped into the car's sound system to enhance the driving experience. I suppose one could call this auditory "information," but it doesn't inform one of anything.[10] When falsification is offered as a remedy for abstraction, we have the engineering equivalent of the last, desperate days of the Roman Empire. Powdered mandarins glided about the Senate, ripe for conquest and slaughter. This decadence did not go unnoticed by the surrounding barbarians, and a new chapter of history began. Maybe the skateboarders could serve this barbarian role for us, at least in the realm of automotive engineering.[11]

ZONES OF REACH AND REPRESENTATIONS

When viewing two-dimensional representations, whether photographs, paintings, or screens, we are not able to move around and gain different perspectives on the scene depicted. Recall that it is by moving around that we "extract invariants from the stimulus flux," as Gibson says. When we can't do that, our basic equipment for reality testing is inoperative.

Further, we normally orient ourselves in our physical environment according to an axis of proximity and distance, and this basic orientation is not available when the world appears through mediating representations.

According to Alfred Schutz, the spatial categories we employ in everyday life arise from our embodiment. A person is "interested above all in that sector of his everyday world which lies within his reach and which arranges itself spatially and temporally around him as its center." Relative to this center, one carves up the surrounding world at its egocentric joints: right, left, above, below, in front of, behind, near, far. The world within "actual reach" is basically oriented according to proximity and distance. This reachable world "embraces not only actually perceived objects but also objects that can be perceived through attentive advertence."[12] Thus it includes, for example, things behind you that are close but currently out of sight. The content of this sector is subject to constant change, due to the fact that we move around.

This idea of orientation around a bodily center helps us to see how the attentional environment that has emerged in contemporary culture is novel and somehow centerless. Recall that the basic concept at the root of attention is selection: we pick something out from the flux of the available. But as our experience comes to be ever more mediated by representations, which remove us from whatever situation we inhabit directly, as embodied beings who *do* things, it is hard to say what the principle of selection is. I can take a virtual tour of the Forbidden City in Beijing, or of the deepest underwater caverns, nearly as easily as I glance across the room. Every foreign wonder, hidden place, and obscure subculture is immediately available to my idle curiosity; they are lumped together into a uniform distancelessness that revolves around me.

But where am I? There doesn't seem to be any nonarbitrary basis on which I can draw a horizon around myself—a zone of relevance—by which I might take my bearings and get oriented. When the axis of closer-to-me and farther-from-me is collapsed,

I can be anywhere, and find that I am rarely in any place in particular. To be present with those I share a life with is then one option among many, and likely not the most amusing one at any given moment. More broadly, to compose a coherent life on the principle of disembodied, ungrounded choice would seem to be a daunting task.

Is the mouse-click a kind of agency? This gesture, emblematic of contemporary life, might be seen as a fulfillment of the thinned-out notion of human agency we have signed on to when we conceive action as the autonomous movements of an isolated person who is essentially disengaged from the world. Our current attentional environment is novel, but as we have already begun to investigate with our discussion of Kant, it was prepared by a long intellectual history.

To repeat a formulation I used in the previous chapter, if choosing replaces doing for the mouse-clicking Mouseke-doer, it figures that such a disengaged self should be especially pliable to the "choice architectures" that get installed in public spaces. As we shall see, in the darker precincts of capitalism things are being *designed* to foster disengagement, to the point of inducing a kind of autism.

5

AUTISM AS A DESIGN PRINCIPLE: GAMBLING

When my oldest daughter was a toddler, we had a Leap Frog Learning Table in the house. Each side of the square table presents some sort of electromechanical enticement. There are four bulbous piano keys; a violin-looking thing that is played by moving a slide rigidly located on a track; a transparent cylinder full of beads mounted on an axle such that any attempt, no matter how oblique, makes it rotate; and a booklike thing with two thick plastic pages in it. Turning a page initiates a song that corresponds to a picture on the page. There are three buttons (square, circle, and triangle) that initiate different melodies; a saxophone-looking thing played by pressing a button; a lever that makes a round thing spin around behind its clear protective cover; and a little panel that slides back and forth between two positions. If I remember correctly, sliding the panel toggles certain of the aforementioned items from one set of programmed responses to a second set.

Turning off the Leap Frog Learning Table would produce rage and hysterics in my daughter. My initial thought was that this thing was like crack cocaine for toddlers. But the analogy didn't quite hold up, as the device seemed to provide not just stimulation but the experience of agency (of a sort). By hitting buttons, the toddler can reliably *make something happen*. Imagine the frustration of dealing with the world through the clumsy interface of your own body, as a toddler who has not yet learned to walk; not learned to hold a crayon or work a pair of scissors. An attempt to roll a ball toward your father on the ground is likely to send it flying up toward your own nose instead.

The appeal of the Leap Frog Learning Table for toddlers frustrated with their bodies appears to be similar to the appeal of slot machines for adults frustrated by life. The latter is explained by Natasha Dow Schüll in her deeply disturbing book *Addiction by Design: Machine Gambling in Las Vegas.* The goal for compulsive machine gamblers is not to win money, as one might suppose, and you cannot understand their addiction without keeping this in mind. The goal is to *get in the zone*: the place where "their own actions become indistinguishable from the functioning of the machine. They explain this point as a kind of coincidence between their intentions and the machine's responses."[1] You hit the button and the machine responds every time.

Schüll in fact notes parallels to children's electronic games, and draws on studies of these that explore a certain paradox. The appeal of the games is that they give the player a sense of control. But precisely because she is able to reliably produce an effect (such as an auditory beep), the player loses herself in the machine and enters a state of absorbed automaticity, which would seem to be the opposite of control. This state is in fact more passive than active. Schüll quotes one scholar who writes that the children's games accomplish this by way of their "unique responsiveness," which "amplifies and embellishes the actions of the user in so compelling a way that it disconnects him from others and obliterates a sense of difference from the machine."[2]

How does this kind of merging with a machine resemble, and how does it differ from, the unity of machine and rider that we explored in the case of motorcycling? Both are instances of cognitive extension. Schüll quotes one of her gambler informants saying, "I get to the point where I no longer feel my hand touching the machine." The informant continues: "I feel connected to the machine when I play, like it's an extension of me, as if physically you couldn't separate me from the machine."[3] This sounds like the way a hockey player talks about his stick, or the way a motorcyclist feels his consciousness "running out through the contact patches" of his tires.

We will consider the strange indifference of hard-core gamblers toward the outcome of their bets shortly. Given this indifference, the response of a slot machine is like that of an electronic toy: exact and consistent. Your action of pressing a button produces an effect that aligns perfectly with your will, because your will has been channeled into the spare, binary affordances provided by the buttons: press or don't press. You give yourself over to the logic of the machine and are rewarded by a feeling of efficacy. That is, you lose yourself, and thereby gain control.

One difference between this and using a mechanical prosthetic to act in the world (as when we use a hockey stick or a motorcycle) is that the latter preserves variability. Small differences in your action produce differences in outcome—indeed exaggerated differences, if the point of the prosthetic is to amplify your actions (as in a hundred-miles-per-hour slap shot, or the minute steering inputs that result in a quick lane change at highway speed). Variations in how you hit the button on a Leap Frog Learning Table or a slot machine do not similarly produce variations in the effect you produce. There *is* a closed loop between your action and the effect that you perceive, but the bandwidth of variability has been collapsed to the point that it can no longer be said that through your actions you are "extracting invariants from the stimulus flux," to borrow Gibson's phrase. You are neither learning something about the world, as the blind man does with his cane, nor acquiring something that could properly be called a skill. Rather, you are acting within the perception-action circuits encoded in the narrow affordances of the game, learned in a few trials. This is a kind of autistic pseudo-action, based on exact repetition, and the feeling of efficacy that it offers evidently holds great appeal.

Schüll refers to a concept called perfect contingency in the literature of child development, which names a situation of "complete alignment between a given action and the external response to that action, in which distinctions between the two collapse." (I find it confusing to call this perfect contingency because it seems rather the complete absence of contingency.) Early infancy is a bit like this: a state of "seeming merger with the mother's body (and by extension, with the wider environment) that derives from the seamless adaptation of the mother's responses to her infant's needs, wants, and gestures." As the baby matures, his mother becomes less immediately responsive, and "the infant gradually accepts that he does not have magical control over the world and learns to tolerate suspense, unpredictability, and frustration, a critical step toward effectively relating to others."[4]

(We should pause to note that the child would not be helped in taking such a developmental step by watching *Mickey Mouse Clubhouse*, which continues the illusion of magical control over the world and encourages the child to view technology as the great Mommy who will respond seamlessly to his will and keep him insulated from the frustrations of a contingent world.)

After they reach about three months of age, babies come to *prefer* "imperfect contingency," in which "environmental responses are closely yet not perfectly aligned with their own vocal or gestural actions in intensity, affect, or tempo."

But note that

> autistic children are an exception; they remain distressed when an exogenous entity does something that demonstrates vitality of its own, and they are especially intolerant of social contingency, or the unpredictability of another's perspective or intentions. Preferring sameness, repetition, rhythm, and routine, they retreat into circular, self-generated perfect contingencies such as rocking or swinging, or object-based interactions that allow close-to-perfect stimulus-response [circuits] such as bouncing a ball or pressing a button.[5]

In playing at a slot machine or video poker terminal, either you are going to win or you are going to lose. One of Schüll's informants tells her, "I don't care if it *takes* coins, or *pays* coins: the contract is that when I put a new coin in, get five new cards, and press those buttons, I am allowed to *continue*. So it isn't really a gamble at all—in fact, it's one of the few places I'm certain about anything . . . If you can't rely on the machine, then you might as well be in the human world where you have no predictability either."[6] The appeal of machine gambling is apparently tied to an experience of the human world as lacking a basic intelligibility.

Perhaps we are all becoming autistic, in this broad sense. If so, it is not without reason. As the world becomes more confusing, seemingly controlled by vast impersonal forces (e.g., "globalization" or "collateralized debt obligations") that no single individual can fully bring within view; as the normative expectation becomes to land a cubicle job, in which the chain of cause and effect can be quite dispersed and opaque; as home life becomes deskilled (we outsource our cooking to corporations, our house repairs to immigrant guest workers); as the material basis of modern life becomes ever more obscured, and the occasions for skillful action are removed to sites overseas, where things are made; to sites nearby but socially invisible, where things are tended and repaired; and to sites unknown, where elites orchestrate commercial and political forces—when all of this is the case, the experience of individual agency becomes somewhat elusive. The very possibility of seeing a direct effect of your actions in the world, and knowing that these actions are genuinely your own, may come to seem illusory.

Escaping to a zone of autistic pseudo-action has understandable appeal. Precisely because this zone has been sealed off from the world, it is experienced as a zone of efficacy and intelligibility.

Advanced economies are said to be moving away from producing goods or delivering services, in favor of *creating experiences*. This necessarily relies on techniques for attracting and holding attention. (For what is an experience, other than an episode in which one's attention is engaged in some way?) Because our experiences are increasingly manufactured for us, it follows that our attention is increasingly structured by *design*.

The point of the design, often, seems to be to produce experiences of highly channeled pseudo-action that gratify the need to exercise the will, even if only in the merely formal sense of pushing a button, or choosing something from a menu of options.

Perhaps this is what is left to us, given the deep contradiction that we live in: on the one hand, we have the individualist ideal—one is tempted to say the autistic ideal—of the unencumbered self who acts in freedom, and on the other hand we feel beset by insecurities and obscurities that emanate from the collective world. These latter are often technological in nature. We therefore seek out other, *personal* technologies that can give us safe haven: "manufactured certainties," as Schüll puts it, that help us "manage [our] affective states." That is what computer games seem to do for our quasi-autistic cohort of young men; it is what machine gambling does for those who have gone down that particular path. Perhaps such pursuits help us manage the anxiety and depression that come when experiences of genuine agency are scarce, and at the same time we live under a cultural imperative of being autonomous. Escape to the autistic zone, where there are no impediments between your will and its realization, is precisely the remedy that is wanted if your life resembles that of the passive kitten on the carousel of modern life, who is nonetheless exhorted at each rotation to "seize the day!"

As we have seen in the case of *Mickey Mouse Clubhouse*, children are educated into this contradiction from an early age. The Handy Dandy machine presents manufactured certainties, the point of which is to reassure the child that every problem is solvable—if only we allow some other entity to *leap in* on our behalf ("Oh Tootles!") and insulate us from the kind of contingencies that easily lead to frustration. As we saw in our treatment of embodied cognition, these are precisely the contingencies we have to learn and accommodate ourselves to if we are to achieve adult agency and join ourselves to the world, grasped as something independent of the self. The alternative offered in *Mickey Mouse Clubhouse* is to leave the self-containment that comes naturally to the toddler undisturbed, and then *manage* the frustration of living by presenting a limited menu of ready-made solutions. Such management makes those educated into the disengaged mode of living more tractable to the "choice architects" who order our collective lives.

Managing frustration by sidestepping the intractable contingencies of life is a growth industry; the demand for manufactured experiences is met by a growing economy of "affective capitalism," as it has been called. This is usually explained with reference to leisure activities like gambling, playing video games, viewing porn, or taking recreational drugs. But the term could also be applied to some jobs. The anthropologist Caitlin Zaloom worked in the financial futures trading pits in Chicago, and relates what it is like to be a derivatives trader who stares at screens of rapidly shifting data, looking for patterns. In this intense, self-enclosed world, which she compares to a video game, traders engineer "peak experiences of attention" for themselves.[7] Traders get into "the zone" (they actually call it this), a state of total absorption where all else falls away. This is possible only because the messy human realities behind the financial entities they are trading in (for example, people's mortgages) are mediated away by layers of representation and mathematical models, allowing a kind of "control without contact." The models become fascinating in their own right; traders enter deep into their logic and live in the data, rapt in the experience of a growing, intuitive grasp of it. Needless to say, in the years leading up to the financial crisis this quasi-autistic financial game caused massive casualties, but they took place somewhere else, out in the world beyond the screen.

The "gaming industry" appears to be the most self-conscious and sophisticated practitioner of the art of attentional design that is currently establishing itself in the economy of affective capitalism, where our experiences are manufactured for us. Let us therefore take a more sustained look at this art, under the premise that it represents an especially clear case of a trend that is growing, but harder to make out elsewhere, because the role played by attention is not as obvious.

ADDICTION BY DESIGN

Schüll's book is arguably one of the more important works of social science to appear in the last thirty years. I can't here do justice to the richness of its reporting, the scope of its interpretation, or the doggedness with which Schüll goes about excavating the engineered reality of the gambling experience. Quite apart from its interest as an exposé and the revelations it offers about new forms of capitalism that are taking shape, the book pieces together important insights about our cognitive architecture and affective drives. By digging deep into the experience of machine gambling, Schüll arrives at a broader anthropology that illuminates the strange tensions in our nature that could make such an activity appealing.

It is not uncommon for heavy users to stand at a machine for eight or even twelve hours at a stretch, developing blood clots and other medical conditions. Paramedics in Las Vegas dread getting calls from casinos, which usually turn out to be heart attacks. The problem is that when someone collapses, the other gamblers won't get out of the way to let the paramedics do their job; they won't leave their machines. Deafening fire alarms are similarly ignored; there have been incidents where rising floodwaters didn't dislodge them. The gamblers are so absorbed that they become oblivious to their surroundings.

Schüll interviews one woman who makes sure to wear dark clothing when she goes to gamble so it won't show when she urinates on herself. Once a gambler has taken possession of a machine, the thought of leaving it is intolerable, and so the urine-and-feces issue turns out to be a fairly common part of the machine gambling experience.

This is not quite the suave image of James Bond at the blackjack table in Monte Carlo. We see him catching sidelong glances from a circle of mutually posturing players who are intrigued with one another as much as with the game. In such a scene, gambling seems to be merely a setting for the exercise of a certain urbanity. The mix of confidence and abandon displayed in a high-stakes bet is attractive. You hope to be bathed in the sparkling light of Fortune before your rivals. There are occasions for courage and composure; you reveal yourself, in winning and losing both. Or perhaps you have an ulterior end and want to demonstrate to a potential business partner a capacity for sober calculation, a toughness of mind that is unconcerned with social display. In any case, you surely don't *piss in your pants.*

This image of gambling as a rich social practice no longer fits the reality. Schüll writes, "Until the mid-1980s, green-felt table games such as blackjack and craps dominated casino floors while slot machines huddled on the sidelines . . . along hallways or near elevators . . ." By 2003, the president of the American Gaming Association estimated that "over 85 percent of industry profits came from machines." The public relations surrounding the machines presented them as mainstream consumer entertainment, like pinball arcade games. State officials looking for gambling revenue were happy enough to accept the industry's redefinition of itself as "gaming," which helped to remove the taint of moral failing or predation. At the same time, consumers were becoming more accustomed to interacting with screens; this was the time when personal computers and video games were becoming common. Schüll points out that these wider developments helped machine gambling come to seem normal.

At the same time, ownership of casinos passed from organized crime to publicly traded corporations, a move made possible in Nevada by its Corporate Gaming Act. The background checks previously required to buy or build a casino did not apply to shareholders. Schüll writes that the new ease of raising capital led Wall Street to take an active interest, and over the course of the 1990s the Las Vegas strip came to be dominated by corporate megaresorts.

The number of tourists visiting the city increased fourfold from 1980 to 2008, with a corresponding swell in the number of full-time residents working in the service industries. And this is where the story of Las Vegas starts to get interesting, as it was from this local population that corporate interests learned the potential of machine gambling. Tourists are merely "transient players," as the industry calls them, while the local population is made up of "repeat players." The plasticity of our brains is such that it is through repetition that addictions are first established.

Schüll writes that "a full two-thirds of those who reside in metropolitan Las Vegas gamble. Of these . . . two-thirds gamble heavily (defined as twice a week or more, for four hours or longer per session), or moderately (one to four times a month, for up to four hours per session)." They do this mostly at neighborhood casinos that offer easy parking and *child care facilities*. Schüll quotes the slot manager at a venue that is popular among residents as saying, "Our local players are very discriminating. They know what they want, and they're here five to seven days a week."[8]

It is from these "discriminating" players who "know what they want" (and apparently want it bad enough to entrust their children to a casino) that the industry now gets most of its profits. The steady repetition of machine play by locals at the dollar slots yields a better revenue stream than the episodic, high-rolling play of tourists who play at green felt tables. The arrangement, then, is this: you work forty hours a week doing food prep in the bowels of some megaresort, cleaning hotel rooms, or working as a security guard or cocktail waitress or reservations clerk on the strip. Then you spend your leisure time feeding your paycheck into the machines. Schüll writes that nearly 82 percent of local gamblers are members of the "loyalty clubs" that casinos offer, "carrying player cards that document the volume of their play and reward them accordingly" with various trifles. Through their loyalty cards, repeat players are tracked and their behavior is carefully analyzed. Some casinos have facial recognition software that enables a player's favorite machine to call out to her by name if cameras on the casino floor detect that she is headed toward the exit.

At its annual convention in 1999, the gaming industry recognized that Vegas locals represented its most "mature" market, and could be taken as a model for the rest of the nation. The fiscal distress of state governments provided an opportunity for expansion, and indeed the machines are now permitted in forty-one states, and can be seen not just in casinos but in bars, gas stations, bowling alleys, restaurants, truck stops, supermarkets, drugstores, and car washes; this is called "convenience gambling."

But we haven't yet considered *why* people park themselves at these machines and feed their money into them. "The speed is relaxing," one of Schüll's gambler informants tells her. "It's not exactly excitement; it's calm, like a tranquilizer." The fact that there is no substance involved in gambling, as there is in chemical addictions, makes it hard for some to accept that it is a genuine addiction, complete with physical withdrawal symptoms. One of the determining factors in the establishment of any compulsive behavior is the frequency of rewards. The frequency has to keep increasing, as we develop tolerance for any given rate of reward.

The speed of play has been accelerated with some fairly straightforward innovations over the years, such as replacing the mechanical pull handle of slot machines with an electronic push button (which you can rest your hand on constantly), which was followed by the mechanically spinning reels being replaced with a video screen. Once the machines accepted bills (in large denominations), one no longer had to insert coins laboriously into the machine; merely eliminating this fumbling generated a 30 percent increase in the amount of money played. Experienced video poker players (you may have seen one hunched at a terminal at a bar or gas station, waving fingers over a touch screen in a blur that rivals the best typists) can complete up to 1,200 hands per hour; the rate of play on video slots is similar, up from about 300 games per hour a couple of decades ago. If you wager a quarter on a game, you may "win" fifteen cents—this loss registers as a win with flashing lights, which get integrated with the dopamine reward circuits in your brain. What gamblers call "the zone," the industry calls "continuous gaming productivity."

Gaming productivity has three components: the speed of play, its duration, and the amount wagered per cycle. Duration of play has been increased by some design elements aimed at eliminating disruptions. The goal, in industry parlance, is to extend "time-on-device." Before large bins were added to slot machines in the 1960s and 1970s to receive the flood of coins that came with a large win, a gambler who scored a jackpot had to stop playing and wait until a casino floor attendant came over to verify his win and pay him before he could continue. Schüll quotes an industry innovator saying, "This didn't just slow down play, it suggested a kind of closure, an end to the game . . . it tempted the customer to cease the play and walk out the door with his winnings." On the other hand, a hopper full of coins was more likely to be fed back into the machine, so the gambler could "gather the wagering momentum critical to the flow of their play experience." Cashless gambling, in which money has been dematerialized into magnetic swipe cards, has "further helped to overcome impediments to play associated with money insertion."[9] Access to the zone is a function of access to cash, and though Nevada law prohibits the integration of ATM functions into the slot machine itself, other jurisdictions are more forward-looking and allow limitless transfers from the gambler to the casino at the site of play, as long as funds (or credit cards) are available. A company called Global Cash Access calls its device for accomplishing this Stay-n-Play.

But the real progress in productivity came when the industry realized that there was an intimate connection between speed and duration: increasing the speed of play makes the experience more absorbing, and hence also tends to extend the duration of play. As Schüll notes, the gaming industry has embarked on a program that resembles the Taylorist time-and-motion analyses of the early twentieth century, whereby the productivity of factory workers was maximized. The goal was to discover the fastest possible rate at which the assembly line conveyors could move, given the limitations of the human body. Of course, in Las Vegas the object of this kind of scientific management is not a producer, but rather a consumer of the manufactured "zone" experience. Still, productivity must be maximized. Schüll quotes industry insiders who forthrightly articulate the design goal of the machines and of the broader casino environment as one that leads players to play "to extinction." That is, until they have no funds left.

This is done, in part, by making sure the player is as comfortable as possible, so his body becomes unobtrusive. He should be insulated from anything extraneous that could compromise the continuity of the zone experience, through careful design of the machine interface and of the broader casino environment. Total immersion is what the gambler wants, and it is also what the casino wants. The designers of the experience call their solicitude "player-centric design." In this happy harmony of preference satisfaction, consumers are empowered. Give the people what they want.

But the appearance of aligned interests between the player and the casino is sustained only by ignoring the obvious mismatch in their perspectives. Players are seeking an intoxicating experience in the moment—an experience outside of time, as Schüll says, in which the pressures and contingencies of life are suspended—while the industry is soberly grounded in the shared world where the clock continues to tick and the cumulative effect of the player's abandon to the zone may be counted up. When the last nickel is spent and the player emerges from the trance, standing in a puddle of urine and blinking at the rising sun, the collusive relationship is revealed to have been one-sided.

Schüll quotes an industry person who spoke on a panel at a gambling convention devoted to player-centrism. "The more you tweak and customize your machines to fit the player, the more they play to extinction; it translates into a dramatic increase in revenue."[10]

In playing a card game such as blackjack, a game of dice such as craps, or a game based on physical contingencies such as roulette, the odds a player faces are apparent to him (assuming the dice are not loaded, etc.). Early slot machines shared this attribute; there were generally three reels, each with a fixed number of stops with various symbols on them. Lining up the same symbol across all three reels issued in a payout. Eventually the weights and springs of the purely mechanical slot machines gave way to switches and motors, and these in turn gave way to computerized machines, where the odds are controlled by a random number generator. Needless to say, the machines can be programmed for whatever odds the house prefers. But this was true in the old days as well; the odds were determined by the number of stops on a slot machine, and the number of reels that had to line up for the player to get a payout. What is new is that the *apparent* odds that are presented to the player are now subject to manipulation, independently of the actual odds. This is done by displaying machine events that seem to represent the process by which randomness is generated, but are in fact completely divorced from that process. This design element sustains our natural assumption that the game is ruled by lawlike mechanical processes that could be mastered, with enough repetition. But this is an illusion.

Some decades ago, manufacturers tried offering machines with larger reels, able to accommodate more symbols, as well as machines with more reels, but when interacting with these machines, players could plainly see that with more symbols added, their odds of winning had been reduced. We acquire intuitions for grasping probabilities over the course of our development from infancy, as embodied beings who negotiate a stable, orderly world. The breakthrough insight of the gaming industry came when it realized that these intuitions can be manipulated—through "virtual reel mapping."

The reels that are displayed retain the traditional number of stops—eleven winning symbols and eleven blanks, for a total of twenty-two—but the virtual reels (where the odds are really being played out) can have any number of stops, sometimes hundreds. So, without any further manipulation, the odds are exponentially worse than they appear to be. But the real magic occurs because the virtual stops can be "mapped" onto the displayed stops however the designer likes. Far more virtual stops are mapped onto "low-paying or nonpaying blank positions on the actual, physical reel [which is displayed] than to winning positions." It gets better still: via a technique known as clustering, "a disproportionate number of virtual stops are mapped to blank spaces just above or below the jackpot symbols. This ensures that they will appear more often above or below the payline than they would by chance alone, enhancing the 'near miss' sensation among players." This makes them want to continue to play. Schüll cites behavioral psychology's "frustration theory of persistence" and the related theory of "cognitive regret," in which "players circumvent regret at having almost won by immediately playing again." The gaming industry is well informed on this literature, and completely self-conscious in what it is doing.[11]

Poignantly, one of Schüll's gambler informants says, "You get to learn the pattern and just need to get it right." Always on the verge of winning, he is led to believe that he is developing an arcane skill, an intuitive connection to the machine's obscure workings. He is not. The frequency with which he almost wins can be made to increase over the course of his play in a single session, and because players are tracked, it can be made to increase from one session to the next as well, leading to a feeling of growing mastery.

Human beings are exquisitely sensitive to detecting patterns, and this is clearly connected to our drive to *become competent*. As I noted before, Nietzsche said that joy is the feeling of your power increasing. Consider the toddler who is gaining mastery of his body, the child who is learning to catch fly balls, or the adult who is leaning hard through a curve on a motorcycle. In all of these endeavors, we achieve competence by becoming sensitive to the patterns by which the flux of sensual data reveals a stable world. We are able to do this because through our own actions we gain different perspectives on our object; our ability to apprehend reality is intimately bound up with our own agency. Virtual reel mapping exploits these basic elements of our cognitive and affective architecture. An exquisite case of attentional design, it creates an illusion of growing competence *because* it creates the illusion of a stable entity to be known, governed by its own necessities. But in fact the connections between the visual data presented to the player and the machine states where the wins and losses are determined are not only arbitrary, they are contrived to deceive.

The longer odds enjoyed by the house because of these manipulations mean that it can afford to offer occasional multimillion-dollar jackpots, which is key to attracting new gamblers—the ones who have not yet discovered the catatonic pleasures of the zone, and naively dip into machine gambling with the hope of winning. One industry insider says that the megajackpots that arrived with virtual reel mapping were "the primary impetus for the meteoric rise of popularity in slot machines."[12] There must be smaller wins as well, at some optimal frequency; this is called the "reinforcement schedule" in the sort of behavioral conditioning that relies on random reinforcement (as opposed to the "classical conditioning" of Pavlov's famous dog, where strict correlations are established between events). In experiments with rats, random reinforcement (in the form of a dose of cocaine) has been found to be the most powerful way to induce the animals to persist in some behavior (for example pressing a button with their snout), for which they are occasionally rewarded. They will persist so doggedly that they neglect to eat or drink, and so they die. Their instinct for self-preservation has been overridden by something more powerful.

THE DEATH INSTINCT

The more advanced stages of machine gambling addiction are explained by Schüll with recourse to Freud's idea of the death instinct. She quotes a gambler named Maria who says, "The only real control you can have over the end is to make it come faster." This sounds like the peculiar, self-negating agency of suicide, and the analogy is apt (quite apart from the prevalence of actual suicide among gambling addicts, which is higher than for any other addiction). The design script of the machines—to enact "player extinction"—mates up with a deep human tendency, one that I believe is not confined to addicts and suicides.

Freud wrote that the mechanical drive to relive a frustrating and painful event hints at "a compulsion to repeat which overrides the pleasure principle." The pleasure principle—the drive to gratify desire and avoid pain—keeps us in perpetual motion. These are the motivating "life instincts," the basis of selfhood. Yet we have also a more primitive set of instincts "to return to a state of rest, stillness, and peace," as Schüll describes the death instinct. The aim is "to extinguish life's excitations and restore stasis."

Schüll quotes a gambler named Sharon who no longer looks at the cards she is dealt in video poker. "You reach an extreme point where you don't even delude yourself that you're in control of anything but strapping yourself into a machine and staying there until you lose . . . All that stuff that draws you in the beginning—the screen, the choice, the decisions, the skill—is stripped away, and you accept the certainty of chance: *the proof is the zero at the end.*"[13]

Gamblers report their annoyance at winning a jackpot in the wee hours, when they are exhausted and just want to go home. But they are compelled to "zero out" a tension that is sustained by winnings. Schüll writes that "their 'gamble for control' [by this I think she means the quasi-autistic sense of efficacy they enjoy in "the zone" of high-speed play] appears to be underwritten by *a wish to move past the need for control altogether.* From this perspective, the financial losses they sustain while gambling are not merely collateral consequences of their bid for control, but instead, its more profound aim."[14] If you have money left, you are obliged to exercise choice. You may have decided hours ago to abandon yourself to the video poker terminal, but as long as you still have funds available, you are faced with the possibility of acting otherwise: of stopping. Your will is still in play. Similarly, sex addicts report that they often seek out a prostitute not out of sexual desire, but in order to put to rest the *question* of whether they will or will not be with a prostitute today. Once they submit to the compulsion, the question is settled and the will is relieved of its burden. For this person, as for the gambler, the real relief lies in being *spent.* Only then can there be a moment of repose. We might view this as an exhausted response to the heightened burden of self-regulation that we bear in a culture predicated on freedom.

Consider the phenomenon of players jamming a toothpick into the button that initiates a play so that the machine plays itself continuously and the player becomes a mere bystander, watching the credit meter rise and fall (mostly fall). In Australia, an "autoplay" feature has been incorporated into slot machines, to serve the "mature" player who has moved beyond control to pure automaticity, and experiences himself as part of the machine. Such desubjectification does look quite a bit like death.

This might seem exotically pathological, but I can detect something like a death instinct in myself, for example in those times when I slump in front of the TV and watch whatever is served up. It becomes an occasion for self-disgust as soon as I rouse myself from the couch, and is no great source of pleasure while I am in the trance, so why do I do it? I think because the passivity of it is a release from the need for control. As someone who is self-employed, I don't have the jig of a regular job, so the disposition of every hour is a matter of choice, an occasion for reflection and evaluation. Sometimes I just want to stay where I am and watch *Dateline*, because *that's what's next.* Let death come.

There is another reason to regard hard-core gamblers not simply as aberrant, but as showing us something important about the human condition. Their activity is perhaps the dark mirror image of something we recognize as worthwhile: an activity that has no point beyond its own continuance, because it is not a means to some other end. The point of the activity is the activity itself. Talbot Brewer has elaborated such "autotelic" activities and finds that they have a certain structure to them: they are guided by intimations of something valuable that you are trying to bring more fully into view through your activity. In the course of your repeated efforts, you find that what you are aiming at is a moving target, because it reveals itself only in the course of your pursuit. Brewer gives the example of a blues singer who is trying to find just the right phrasing to convey the subtle and complex emotional register she is aiming at in a particular passage of a song. In doing so she is not simply finding the means to express an emotional truth that she is already fully in possession of. Rather, she is finding that emotional truth in the course of singing it this way and then that way.

In Brewer's account of an autotelic activity, there is some reality to be apprehended—for example, emotional truth in the context of a particular song. With gambling machines, the sense of something real to be apprehended is conjured by various manipulations of our capacity for detecting patterns, and this probably contributes to their absorbing nature. But as we have seen, this deception fades away as the gambler gains experience, and he comes to embrace "extinction" as the goal of his activity.[15]

This is indeed alarming, and hard to square with our usual notions of what makes an activity appealing. If we understand it as a pathology in the individual gambler, this has a certain calming effect. And in fact, the gaming industry was quite pleased when the *Diagnostic and Statistical Manual of Mental Disorders* included "pathological gambling" in its menu of psychiatric disorders, as this helped the industry's own efforts to characterize "problem gambling" as the manifestation of predilections in the individual—a preexisting inability to resist internal impulses—that, while certainly regrettable, is characteristic only of a separate class of gamblers, who no doubt would find some other outlet for their self-destructive behavior if slot machines were regulated by a meddling and paternalistic state.

For the industry, the point of characterizing obsessive gambling as a medical condition of the individual is, of course, to distract attention from the fact that conditioning gamblers to play "to extinction" is the design script that animates every aspect of the gambling experience, from the interior design of casinos to the minutiae of the machines' displays to the carefully calibrated frequency of wins. As Schüll says, the addictive quality of the machines is neither a property of the machine itself nor simply a predilection of the individual, but arises from an interaction between our (normal) psychological makeup and the dark arts of attentional design. The plasticity of our neural pathways is such that repetition coupled with random reinforcement issues in addiction. This is the foundation of the business model.

Schüll writes, "While the medicalization of excessive gambling helped somewhat to undermine condemnations of gamblers as weak of will or morally compromised, ultimately it did more to undermine condemnations of gambling vendors as purveyors of a socially and morally corrupting activity."[16] The medicalization of what have previously been considered moral issues is a broader cultural phenomenon. This trajectory is irresistible—who wants to be the last asshole standing, issuing condemnations rather than solicitude? But here we see the sly logic by which democratic nonjudgmentalness gets turned to advantage in unregulated capitalism, with the aid of an expansionary psychiatric establishment. To capital, our moral squeamishness about being "judgmental" smells like opportunity.

THE LIBERTARIAN RESPONSE

Our economic system assumes that individuals are radically responsible for themselves. Maintaining this view requires that we hive off any group of people who fail to live up to the autonomous ideal (problem gamblers, sex addicts, etc.) and designate them pathological. If they have an *internal* defect, then there is no urgent reason to criticize external forces (for example, slot machines in convenience stores; porn that is accessible on your mobile device) that contribute to their lack of self-command. The creeping saturation of life by hyperpalatable stimuli remains beneath the threshold of concern if we repeat often enough the mantra that "government interference" is bad for "the economy." It would certainly be bad for the bottom line of some particular people.

In a segment on gambling addiction on the news program *60 Minutes*, Lesley Stahl interviewed Ed Rendell, then governor of Pennsylvania, who had been especially aggressive in courting the gaming industry. He began by insisting that "the upside" to having casinos in the state is significantly bigger than "the downside." This language of cost-benefit analysis might be fitting if there were a single entity that were subject to both the upside and the downside. But of course, the benefit to the state is due to a *transfer* of wealth, and as taxes go it is a highly regressive one, hitting lower-income people hardest. It also cannibalizes other forms of taxation such as sales taxes, since money spent on gambling isn't spent on other things. But the interview got more interesting than you might have expected from these initial gambits. Rendell spoke of the timeless appeal of gambling, and was ready with some references to gambling on the banks of the Tigris and Euphrates, the cradle of civilization. He insisted that people are going to gamble anyway, so they may as well do it in Pennsylvania. As for problem gamblers, "anyone who has that bent is going to lose their money anyway." Pressed by Stahl on some glaring points of weakness in his position (for example, if you only have to drive across town to gamble and your local newspaper is blanketed with full-page ads and promotions for casinos, as happened in Pennsylvania, this is quite different from having to travel to Atlantic City), the governor became quite angry, turning from Stahl and fulminating directly into the camera.

Rendell seemed to take ideological offense at the line of questioning he was being subjected to. The interview broke off when he became upset, but Stahl reported that his bottom line was that "people should be allowed to make their own decisions." Who could be against that? But the effectual truth of this kind of libertarian autonomy talk is to guarantee that some nontrivial portion of the citizens of Pennsylvania will become . . . how shall we put this . . . loyal gamers.

This illustrates a broader point. We abstain on principle from condemning activities that leave one compromised and degraded, because we fear that disapproval of the activity would be paternalistic toward those who engage in it. We also abstain from affirming some substantive picture of human flourishing, for fear of imposing our values on others. This gives us a pleasant feeling: we have succeeded in not being paternalistic or presumptuous. The priority we give to avoiding *these* vices in particular is rooted in our respect for persons, understood as autonomous. "People should be allowed to make their own decisions."

Liberal agnosticism about the good life has some compelling historical reasons behind it. It is a mind-set that was consciously cultivated as an antidote to the religious wars of centuries ago, when people slaughtered one another over ultimate differences. After World War II, revulsion with totalitarian regimes of the right and left made us redouble our liberal commitment to neutrality. But this stance is maladaptive in the context of twenty-first-century capitalism because, if you live in the West and aren't caught up in battles between Sunnis and Shiites, for example, and if we also put aside the risk of extraordinary lethal events like terrorist attacks in Western countries, then the everyday threats to your well-being no longer come from an ideological rival or a theological threat to the liberal secular order. They are native to that order.

Those with a material interest in doing so have learned to speak autonomy talk, and to tap into the deep psychology of autonomy in ways that lead to its opposite. Further, as the governor's speech illustrates, our original liberal principle of value agnosticism neutralizes our critical energies.

If we have no robust and demanding picture of what a good life would look like, then we are unable to articulate any detailed criticism of the particular sort of falling away from a good life that something like machine gambling represents. We are therefore unable to offer any rationale for regulation that would go beyond narrow economic considerations.[17] We take the "preferences" of the individual to be sacred, the mysterious welling up of his authentic self, and therefore unavailable for rational scrutiny. The fact that these preferences are the object of billion-dollar, scientifically informed efforts of manipulation doesn't square with the picture of the choosing self assumed in the idea of a "free market." It is a fact without a noisy partisan, so our attention is easily diverted from it. Further, by keeping his gaze away from such facts, the liberal/libertarian keeps his own soul pure, lest he commit the sin of recommending to others some substantive ideal, one that will necessarily be controversial. But outside his garden wall there are wolves preying on the townspeople. In our current historical circumstances, his liberal purity amounts to a lack of public-spiritedness.

Even on its own terms, Governor Rendell's libertarian reflex doesn't seem quite apt in the context of machine gambling. Clearly he is a lover of freedom. If asked "freedom from what?" he would presumably answer, "the government." But this begins to seem a strangely eighteenth-century answer, when you think about a typical day in the twenty-first century. You may find yourself on the telephone, caught in an unwinnable battle with TRW or some other credit rating bureau because they made a mistake—perhaps a very consequential one—in administering your credit history. While on hold you get on the computer and try to figure out a recurring, unexplained charge on your Verizon cellular bill. Our founding republican spirit of "No taxation without representation" and "Don't tread on me" is laudable, but must be directed to the proper offshore entity. Libertarians are confused because, unlike King James I, Verizon doesn't make a straightforward assertion of sovereignty. Instead, it wraps you up in the embrace of rational-looking bureaucratic irrationality. While in this embrace ("Your call is important to us"), one catches a distinct odor of bad faith and begins to suspect that the irrationality one is battling is not due to a system error, but part of the business plan. Perhaps the time one spends on hold is not due to "unusually heavy call volume," but is rather calibrated to persuade a certain percentage of callers not to persist. Those who do persist are subject to recorded advertisements for other Verizon services, for a period equally well calibrated, to keep the caller just this side of the threshold beyond which he decides to go Unabomber.

Verizon makes no reference to the divine right of kings; the authority it acts under cover of is that of "contract"—putatively, an agreement you entered into freely, with full information and plenty of alternatives. If you fail to pay the mysterious charges, you will not face an armed force (at least, not right away), but you will take a further hit to your credit score.

I appreciate the freedom-loving, government-hating spirit of libertarians, but I think they take too narrow and old-fashioned a view of the thing they hate—of the settings in which the individual is subject to various kinds of rule. Capital is concentrated to the point that it operates in quasi-governmental ways, abetted by ever more powerful information technology. Arguably, one of the most important functions of the (actual, elected) government, now, is precisely to restrain and regulate the explosion of unaccountable governmentality in our dealings with outsized commercial enterprises. I am happy to pay the IRS my share, if the funds it collects will help the government maintain its monopoly on coercive power, not least by regulating commerce. I want the Federal Trade Commission to fight TRW on my behalf. It is possible, then, to make a libertarian argument for proactive government, and in the case of machine gambling, this seems worth doing.

On the one hand, the defense of machine gambling by libertarians, as well as the industry's own portrayal of it as free-spirited gaming, assumes an autonomous subject capable of acting in his own self-interest. On the other hand, the machines and every aspect of the casino environment are deliberately engineered to induce people to play "to extinction."

The success of those engineering efforts offers good evidence that we are indeed situated beings, formed in very consequential ways by our interactions with our environment. The business model of the gambling industry is based on this recognition. To merely complain that the industry's manipulations violate our ideal of autonomy is to offer a hapless liberal critique, the haplessness of which consists in the fact that it is based on an unrealistic anthropology: the same picture of the autonomous self that the spokespeople of gambling offer, when they are not acting on the more realistic anthropology that actually guides their efforts.

The more effective defense would consist of a good offense: a positive account of action in its full human context, in which the actor is in touch with the world and other people, in comparison with which the autistic pseudo-autonomy of manufactured experiences is revealed as a pale substitute. My hope is that the accounts I offer in this book of ordinary activities such as that of the short-order cook, the hockey player, and the motorcycle rider (and soon the glassmaker and the organ maker), will help to provide some concrete images that can serve this role. They are images of what I take to be well-ordered ecologies of attention and action, the sort that can support some low-to-the-ground, perfectly attainable moments of human flourishing.

To clear the way further for this positive account of attention, we need to understand better the anthropology of the autonomous self. I have suggested that it underlies the apologetics of the gambling industry, but of course it didn't begin that way. It has its origins in the most august and serious-minded efforts of the Enlightenment. In the interlude that follows, we will investigate it in its original context, the better to see how it fits or fails to fit our current circumstances.

INTERLUDE

A BRIEF HISTORY OF FREEDOM

When we talk about freedom, what we are keen to be free *from* is a moving target. Today's conservatives, if they have an intellectual bent, often refer to themselves as "classical liberals." The term is apt; the view of freedom that they generally cherish is one that was articulated at the founding of modern liberalism by John Locke and others. If you visit Thomas Jefferson's home at Monticello, you will see a prominent portrait of Locke in the parlor, and his *Two Treatises of Government* is echoed at various points in the Declaration of Independence. For the founding generation, the thing we needed to be free from was clear: the arbitrary exercise of coercive power by the political sovereign, who lived in England.

At the close of the previous chapter I turned this "classical" mantle into a criticism, suggesting libertarians have an outdated view of where the threats to freedom lie. This may have seemed an indulgent digression into politics in a book ostensibly about attention. But in fact the birth of liberalism is a crucial moment for our inquiry, because Locke fleshed out the idea of freedom in a way that was necessary for his political arguments, but also resonated far beyond politics, and continues to inform the ideal of autonomy that has become second nature for us. Locke's redescription of politics required a redescription of the human being, and of our basic situation in the world. Ultimately it required a new account of how we *apprehend* the world.

To anticipate:

- We are enjoined to be free from authority—both the kind that is nakedly coercive and the kind that operates through claims to knowledge. If we are to get free of the latter, we cannot rely on the testimony of others.
- The positive idea that emerges, by subtraction, is that freedom amounts to radical self-responsibility. This is both a political principle and an epistemic one.
- We achieve this, ultimately, by relocating the standards for truth from outside to inside ourselves. Reality is not self-revealing; we can know it only by constructing *representations* of it.
- Attention is thus demoted. Attention is the faculty through which we encounter the world directly. If such an encounter isn't possible, then attention has no official role to play.

Let's step back for a moment. In this book I am picking out a few topographical features of modern culture, and suggesting that we see them as part of a larger landscape. Like trees in the foreground, we have various polemics about our novel digital landscape; the larger forest consists of a set of assumptions about how our minds work that we have inherited from the Enlightenment. Of course, these were not originally assumptions, but well-articulated assertions. As such, they were addressed to somebody; they were part of a conversation. In recovering this historical context, we see that the conversation didn't start out as a serene inquiry into how our minds work. It began as a quarrel about politics.

The quarrel was "won," as a historical fact, by the party that was directed by a single master principle: to *liberate*—whether from the ancien régime, ecclesiastical authority, or Aristotelian metaphysics. That is why the term "liberalism" is useful for characterizing the big metaphysical and anthropological picture that was established in these revolutionary centuries. But what does intellectual history, looking back three hundred years, have to do with the current crisis of attention? Quite simply, the experience of *attending* to something isn't easily made sense of within the Enlightenment picture. To see our way through our current predicament, we need a good account of how attention works. And to get *that*, we first need to become more self-conscious about this intellectual inheritance, and hold it up to scrutiny.

Doing so will help us see an underlying unity in the features of contemporary life we have examined thus far. We have considered the problem of mental fragmentation and arbitrariness that results when our contact with the world is mediated by representations: representations collapse the basic axis of proximity and distance by which an embodied being orients in the world and draws a horizon of relevance around itself. We noted the prominence of a design philosophy that severs the bonds between action and perception, as in contemporary automobiles that insulate us from the sensorimotor contingencies by which an embodied being normally grasps reality. The case of machine gambling gave us a heightened example of this kind of abstraction, and made clear how such a design philosophy can be turned to especially disturbing purposes in the darker precincts of "affective capitalism," where our experiences are manufactured for us. We saw that the point of these experiences is often to provide a quasi-autistic escape from the frustrations of life, and that they are especially attractive in a world that lacks a basic intelligibility because it seems to be ordered by "vast impersonal forces" that are difficult to bring within view on a first-person, human scale. I argued that all of this tends to sculpt a certain kind of contemporary self, a fragile one whose freedom and dignity *depend* on its being insulated from contingency, and who tends to view technology as magic for accomplishing this. For such a self, choosing from a menu of options replaces the kind of adult agency that grapples with things in an unfiltered way. Finally, I argued that such a choosy self is especially pliable to the "choice architectures" that get installed on our behalf by various functionaries of psychological adjustment.

These features of our world are hard to criticize because, though they may be appalling once described in the way that I have, they are intimately connected to our defining virtues as modern Western people. I have already suggested that much of what I have described can be understood as a cultural working out of Kant's ideal of autonomy. Now I want to go back further, behind Kant, and investigate some earlier moments in the Enlightenment when we first got set on our trajectory. This is likely to cut close to home, as these earlier moments saw the articulation of those political principles that we rightly cherish. I believe they are robust enough that we may continue to cherish them while taking a cold-eyed look at the way they reverberate out from politics to inform wide swaths of culture, in ways that may no longer be well suited to our circumstances.

It is instructive to regard our current landscape, and the ideal self who inhabits it, as the sedimented result of a history of forgotten polemics, the common feature of which is that they have been animated by the will to liberate. Self-understanding, then, requires digging down into the history of philosophical thinking, for it is in these quarrels that the sediments have gotten deposited. The point isn't to reach bedrock—some foundational, ahistorical self—but rather to do like a geologist and get a clear sectional view of the strata. If we could accomplish this, I think it would help us see the topography of current experience a little differently.

THE COUNTERFACTUAL ORIGINS OF LIBERALISM

For John Locke, the main threat against which it was necessary to assert freedom was the arbitrary exercise of coercive power by the political sovereign. The political theory that prevailed at the time legitimized such power by positing a fundamental difference in kind between the sovereign and everyone else. Various arguments tied monarchy to God's will: the sovereign was God's representative on earth, or there was a nestled order such that child is to parent as citizen is to sovereign, and sovereign is to God. Locke's strategy, however sincere (and scholars disagree on this), is to offer a theological argument of his own: God is so much greater than man, the difference is so unfathomable, that this relation mocks any attempt by one man to claim godlike coercive power over another.[1] We are all equal in our smallness before God. Therefore our natural estate is one of freedom in relation to one another.

Locke spells this out further: once upon a time we lived in a "state of nature," the defining feature of which was the absence of some recognized authority, a third party to arbitrate disputes. At some points in the *Two Treatises* this appears to be a historical claim about how we once lived; at other points it is a conceptual device to describe the moral relations that obtain between persons who have not consented to a common government. In this state, it is merely the dictates of one's own reason that one obeys—there is no such thing as "authority." The problem is that this tends to become a "war of all against all," as Thomas Hobbes had put it. Political society is instituted in a decisive moment when people give their consent to abide by a common judge in whom they invest authority, at which point they acquire political rights and responsibilities. The issue of consent is key; this is the source of the legitimacy of all authority, and of the rights one retains against that authority.

We may allow ourselves to wonder, *when* does this all-important act of consent happen? I was born into a society that was already up and running, and isn't this the case for almost all of us? Maybe I give my consent to the regime tacitly, for example by walking on the public roads. But I don't have much choice in this, do I? If I veer off the public road and try to bushwhack my way over land, I will quickly encounter NO TRESPASSING signs. *Other people* got here first. Locke's theory of legitimate authority founded on consent describes not the normal course of things but a hypothetical moment of political founding. It is not the founding moment of any actual revolution, but of a fable in which there is no already existing society and the land is unclaimed. At the foundation of our political anthropology is a creature who comes into existence in a moment of free deliberation (shall I consent to this arrangement?) that occurs in a present unconditioned by the past. The freedom of the liberal self is the freedom of newness and isolation. Locke's state-of-nature thought experiment is explicitly counterfactual. Its premise is that "you can understand man and his moral and practical endowments only in isolation from the settings in which he might realize those endowments or, much less, be endowed with them in the first place," as Matt Feeney puts it.[2] The liberal self is not situated.

FREEDOM AS SELF-RESPONSIBILITY

Locke's concern with illegitimate authority extends beyond the kind that is nakedly coercive to the kind that operates through claims to knowledge. His political project is thus tied to an epistemological one. The two are of a piece, because "he is certainly the most subjected, the most enslaved, who is so in his Understanding." Locke does some of his most consequential liberating in his *Essay Concerning Human Understanding.*

Charles Taylor points out that "the whole *Essay* is directed against those who would control others by specious principles supposedly beyond question."[3] These are the priests and the "schoolmen," those carriers of an ossified Aristotelian tradition. The Reformation notwithstanding, political authority and ecclesiastical authority remained very much entwined and codependent in Locke's time.

Political freedom requires intellectual independence, then. Locke takes this further. Following Descartes, he calls on us to be free from established custom and received opinions, indeed from other people altogether, taken as authorities. "We may as rationally hope to see with other Mens Eyes, as to know by other Mens Understandings . . . The floating of other Mens Opinions in our brains makes us not a jot more knowing, though they happen to be true."[4]

The project for political freedom thus shades into something more expansive: We should aspire to a kind of epistemic self-responsibility. I myself should be the source of all my knowledge; otherwise it is not knowledge. Such self-responsibility is the positive image of freedom that emerges by subtraction, when you pursue far enough the negative goal of being free from authority.[5]

But this self-responsibility brings with it a certain anxiety: If I have to stand on my own two feet, epistemically, this provokes me to wonder, how can I be sure that my knowledge really *is* knowledge? An intransigent stance against the testimony of others leads to the problem of skepticism.

How do we know some evil genius hasn't deceived us? Even our own senses lead us astray, for example in optical illusions. Descartes takes the very existence of an external world as a legitimate problem for philosophy to worry about. In his search for *certainty*—for a foundation for knowledge that would be impervious to skeptical challenge—it occurs to him that the experience of thinking ("I am thinking") is beyond doubt. If I am thinking, *I* must exist. This is the secure beginning point that must serve as the foundation for knowledge altogether. What we need, then, are rules for the conduct of the mind, which we can follow from this secure beginning to build up certain knowledge. It is not the content of our thinking that matters now, but how we arrive at that content. To repeat Locke's formula, "The floating of other Mens Opinions in our brains makes us not a jot more knowing, though they happen to be true." This entails a new conception of what it means to be rational. The standard for rationality is no longer substantive, but procedural, as Taylor points out. And this means that the standard for truth is relocated: it is no longer found out in the world, but inside our own head; it is a function of our mental procedures.[6]

Attention is therefore demoted. Or rather, it is redirected. Not by fastening on objects in the world does it help us grasp reality, but by being directed to our own processes of thinking, and making *them* the object of scrutiny. What it means to know now is not to encounter the world directly (thinking that you have done so is always subject to skeptical challenge), but to construct a mental *representation* of the world. Another early modern thinker, Giambattista Vico, summed up this view very succinctly: *We know only what we make.*[7]

TRUTH AS REPRESENTATION

Vico's motto captures pretty well the revolution in science carried out by Galileo and Newton. Natural science became for the first time *mathematical*, relying on mental representations based on idealizations such as the perfect vacuum, the frictionless surface, the point mass, and the perfectly elastic collision. What this amounts to, Martin Heidegger says, is "a projection of thingness which, as it were, skips over the things."[8]

One way to state the conviction that all of these Enlightenment figures shared is that *reality is not self-revealing.* The way it shows up in ordinary experience is not to be taken seriously. For example, we see a blue dress, but "blue" isn't in the dress; it is a mental state. Descartes and Locke both insisted on a distinction between "primary qualities," which are properties of things themselves, and "secondary qualities," which are a function of our own perceptual apparatus. The true description of the dress would refrain from invoking the latter sort of property, and say not that it is blue but that its fabric reflects light of a certain wavelength (as we would now say), which we see as blue. We are to take a detached stance toward our own experience, and subject it to critical analysis from a perspective that isn't infected with our own subjectivity.[9]

Let us pause for a moment to let the weirdness of all this sink in. Notice that we have moved from an argument about the illegitimacy of particular political authorities in the seventeenth and eighteenth centuries, to the illegitimacy of the authority of other people in general, to the illegitimacy of the authority of our own experience.

In telling the story of the Enlightenment in this sequence, I want to suggest that the last stage, the somewhat anxious preoccupation with epistemology, grows out of the enlighteners' political project of liberation. Their organizing posture against authority compelled the enlighteners to theorize the human person in isolation—abstracted from any pragmatic setting in which he might rely on the testimony of others, or indeed on his own common sense as someone who has learned how to handle things. It is probably true that reality is not self-revealing *to the detached bystander.* But that is because, as I have argued in Part I of this book, it is by "having to do with" things that we grasp them—not simply as subjects, but as agents. We do this every day, finding our way through a world that we share with others. The passive, isolated observer who is posited as the beginning point for the

Cartesian/Lockean account of knowledge is a person who has been shorn of those practical and social endowments by which we apprehend the world. If such a creature actually existed, we can well imagine that he would be gripped by the question of how he could know anything. For this person, the "thing in itself" would indeed be an inaccessible mystery.

This is our condition, according to Enlightenment epistemology. Today's mainstream cognitive psychology inherits this view, and proceeds on the assumption that *representation* is the fundamental process by which we apprehend the world. This process happens entirely within the bounds of the skull; we may as well be brains in vats. The new ideas about embodied perception and cognitive extension, which connect thinking to doing, pose a radical challenge to this entire picture. I believe one reason the new ideas are being resisted (in addition to the usual sociological reasons why disciplines resist new ideas) is that, as I have just demonstrated, the origins of modern epistemology are intimately bound up with the origins of our moral-political order.

What is at risk, when we start revisiting the question of how we encounter things, is the whole chain of forgotten polemics by which a very partial view of the human person got installed in our self-understanding: the anthropology of modern liberalism.

According to that understanding, other people play an entirely negative role in our efforts to grasp reality and to achieve intellectual independence. In Part II, "Other People," I will argue precisely the opposite.

PART II

OTHER PEOPLE

6

ON BEING LED OUT

The word "education" comes from a Latin root that means "to lead out." To be educated is to be led out of oneself, perhaps. Consider the experience of learning a foreign language, beautifully described by Iris Murdoch:

> If I am learning, for instance, Russian, I am confronted by an authoritative structure which commands my respect. The task is difficult and the goal is distant and perhaps never entirely attainable. My work is a progressive revelation of something which exists independently of me. Attention is rewarded by a knowledge of reality. Love of Russian leads me away from myself towards something alien to me, something which my consciousness cannot take over, swallow up, deny or make unreal.[1]

To learn Russian is to acquire new powers of expression, and probably of thinking too. One acquires the ability to act in settings that would otherwise be mystifying. Our fixation on autonomy clouds our understanding of such development because the skills one exercises in any impressive human performance are built up through submission—to "authoritative structures," to use Murdoch's phrase. Such structures provide those ecologies of attention in which minds may become powerful, and achieve genuine independence. In this chapter, then, I want to explore the possibility that there is a tension between the ideal of autonomy and education.

This suggestion may go down hard, as autonomy is arguably the central totem of modern life. It hovers about our concepts of individuality, creativity, and any number of other terms that convey the existential heroism we're expected to live up to on a daily basis. It is an idea that we moderns have made our dignity hinge on.

EMPOWERMENT THROUGH SUBMISSION

Consider another example: the process of becoming a musician. This necessarily involves learning to play a particular instrument, subjecting one's fingers to the discipline of frets or keys. The musician's power of expression is founded upon a prior obedience. To what? To her teacher, perhaps, but this isn't the main thing—there is such a thing as the self-taught musician. Her obedience rather is to the mechanical realities of her instrument, which in turn answer to certain natural necessities of music that can be expressed mathematically. For example, halving the length of a string under a given tension raises its pitch by an octave. These facts do not arise from the human will, and there is no altering them. The education of the musician sheds light on the basic character of human agency, namely that it arises only within concrete limits. As the example of learning Russian illustrates, these limits need not be physical; the important thing is rather that they are external to the self.

There are yet other layers to the musician's obedience: she plays a prior composition. Or she may improvise, but does so within given melodic forms. These are not natural givens, but rather cultural ones—the mixolydian scale, or an evening raga. At a broader level of musicality, she plays within a genre. It may be hard bop or West Coast cool, Hindustani or Karnataka, or some synthesis of her own, but not invention ex nihilo. To be sure, if one inquires historically, one finds that cultural forms *are* products of human will as exercised in the past; someone had to invent the mixolydian scale. But from the standpoint of any particular individual in the present, they are experienced as a horizon of possibility that has already been set (they are an "inheritance," to anticipate the theme of Part III). Indeed, contingent cultural forms have the character of necessity for most people, us nongeniuses.

Once, while listening to the bluegrass guitarist Tony Rice in concert, I had the thought *He can do whatever he wants*. Such was his complete command of his instrument. Yet "freedom" doesn't seem quite the right concept to capture this expressive power, if by that term we mean an untutored exercise of the will. His freedom, if that's what it was, was artistically compelling because of the musical ideas it was in the service of. These ideas were his own, but not *simply* his own. His expressive power was born of artistic formation.

The kind of collaborative improvisation that takes place among musicians in bluegrass, jazz, or classical Indian music is a good example of what I mean by an ecology of attention. It is mutually adaptive. The improvisation is possible because all parties are attending to one another. It is fruitful only because they are also steeped in forms; the history of their art has become the genetic material, the constitutive fiber, of their own creativity. A master jazz musician quotes from *The Real Book* with the same ease that a master preacher does from the gospels, and the allusion is gotten. It may be taken up and commented upon by the other players; it may be pushed forward toward possibilities that hadn't existed moments before, as they come into being only through the improvisation itself. One must be alert, opportunistic. As in ecology, that is how new forms arise.

Note that worries about "conformity" versus "individuality" are simply put aside in the account of creativity I have just sketched. More strongly: membership in a community is a prerequisite to creativity. What it means to learn Russian is to become part of the community of Russian speakers, without whom there would be no such thing as "Russian." Likewise with bluegrass. These communities and aesthetic traditions provide a kind of cultural jig, within which our energies get ordered.

I think this is obvious. Yet to emphasize community in this way is to stand athwart one of the main veins of the American creed, our individualism. We are Cartesians without having to read Descartes, as Tocqueville famously said. Descartes began his inquiries by putting aside all supposed knowledge received from "example or custom" in order to "reform my own thoughts and to build upon a foundation which is completely my own."[2] On the Cartesian view, being rational requires freeing your mind of any taint of authority of the sort that operates in communities. Kant concurs: Enlightenment is "man's emergence from his self-incurred immaturity . . . [This immaturity consists not in a] lack of understanding, but lack of resolution and courage to use [one's own understanding] without the guidance of another." Further, "*laziness and cowardice* are the reasons why such a large portion of men . . . remain immature for life."[3]

As this language suggests, epistemic individualism is a *moral* ideal, at least as much as it is a doctrine about how we acquire knowledge. It is closely related to the ideal of "authenticity" that shows up throughout American letters. "Society everywhere is in conspiracy against the manhood of every one of its members," Emerson wrote in his essay "Self-Reliance."[4] Walt Whitman's democratic hero "walks at his ease through and out of that custom or precedent or authority that suits him not." Whitman goes on: "You shall no longer take things at second or third hand, nor look through the eyes of the dead . . . nor feed on the specters in books." To live authentically, Norman Mailer would write a century later, one has to "divorce oneself from society, to exist without roots, to set out on that uncharted journey into the rebellious imperatives of the self."[5]

In his masterful book *The Masterless*, Wilfred McClay writes that after the experience of totalitarianism (largely as conveyed by émigré scholars), American intellectuals in the 1950s were alert to any threat against the individual, and found plenty such at home. Mailer was not alone in seeing little difference between a quick death at the hands of the state and a slow death by conformity. For this generation, McClay writes, "the fantasy of devouring totalism and the fantasy of an unencumbered self went together, standing in symbiosis, testimony to a continuing reliance upon an uncertain notion of individual autonomy—and an even more unsteady conception of the grounds (if any) for genuine social connectedness."[6]

The uncertain notion of individual autonomy that McClay refers to would seem to be one that regards autonomy as the simple opposite of heteronomy. From the Jacksonian to the Beat era, *other people* have often appeared to the American as a disfiguring source of heteronomy. In a culture predicated on this autonomy-heteronomy distinction, it is difficult to think clearly about attention—the faculty that joins us to the world—because everything located outside your head is regarded as a potential source of unfreedom, and therefore a threat to the self. This makes education a tricky matter.

MAKING GLASS: JOINT ATTENTION IN ACTION

I once watched a group of three glassmakers work together. Peter Houk is director of the MIT Glass Lab and one of the leading glass-blowers in the United States. Erik Demain is a professor of computer science at MIT; Martin Demain, his father, is both an artist-in-residence at MIT and the coauthor of some hundred-odd scientific papers, most of them with his son. The three of them get together to make glass as a team several times per week. Their collaboration appears to have begun out of intellectual curiosity, and then grew into something more purposive and consuming—so much so that it threatens to take over their academic careers. They now find themselves advancing the state of the art of glassmaking for its own sake. In doing so, they are self-consciously participating in an ancient art that goes back to the times of the Egyptian pharaohs.

I watched them design and then fabricate a piece of "cane" (like a candy cane or barber shop pole) about fifteen feet long, starting from molten blobs of different-colored glass that they stuck together. Being MIT guys, they first designed the cross section of the cane on a computer. Being experienced, they were able to anticipate the transformations of a cross section as it gets twisted and elongated. And, vice versa, they were able to work backward from a desired effect in the cane to a cross-sectional shape that would produce such an effect once twisted and elongated, in what mathematicians call a "screw transform." They developed a computer program to enhance their ability to visualize the process, and to help novices see these things too, at the design stage.

The first thing that hits you in the "hot shop" is the sheer beauty of molten glass. Its colors are various, depending on its temperature and chemistry. The air surrounding a blob of glass seems to become liquid, shimmering with heat waves that trail off in eddies.

Peter, Erik, and Martin gather from the furnace blobs of glass of various sizes and shapes (therefore having different thermal masses and weird thermal gradients within them) and maintain the fluid, molten state of these blobs by frequently inserting them into another furnace, the "glory hole." Timing is everything. Sometimes they will cool part of the surface of a blob by dipping one end of it in water, or give a superficial heating to the skin with a propane torch. In an email Peter Houk writes, "It's important when working on a complex piece to pay close attention to how the plan is evolving and to be able, as a team, to shift directions when necessary, often very fast. Communication is very important. Some moments require moving even faster than verbal communication can allow for."

The glass has a certain urgency to it, but there was no hint of panic in this team. Indeed it was striking how calm they were as they moved around the shop in concert. Houk believes this kind of cooperation is "one of the key things our program is teaching students at MIT." Being MIT students, they often want to reduce the process to a set of formulas describing heat transfer, viscosity, and the like. But the morphing of molten glass—its drooping, turning, and solidifying—is something you have to feel from the end of a rod. Houk says you can't really see the heat transfer that is pertinent to your plan of manipulations; only by actually manipulating the glass does it convey its current state and likely trajectory.

For a complex piece, these manipulations require more than one pair of hands, each attuned to the current state of their part of the whole. They also have to be attuned to what is taking shape in the hands of their collaborators.

Houk is generally the "gaffer": the team leader who sits at the bench and is responsible for communicating the plan to his (tenured) assistants. He and the piece he is working on are the center of attention. From that position he conducts a kind of group dance that has to be adaptively fluid, or molten, because the glass itself has these qualities. He tells me:

"Different gaffers have different styles as far as how verbal they are in communicating their plan to their team before starting a project and during the making of a hot piece. Some gaffers, like the famous Venetian glassblower Lino Tagliapietra, barely say anything at all, even at the outset. Maybe just a few words about how the process will start out, and rarely any drawings of the finished product. I've watched him work many times, and his process reminds me a little of how Miles Davis worked with his bands: some structure to work with, but not too much information and then the rest improvisation within a fairly structured system. His assistants have to be able to read what is going on from nonverbal cues and by looking at what the glass is doing. That's why he has had the same team for fifteen-plus years. It's very typical in the Venetian tradition for a master and his first and second assistant to stay together for an entire career, and watching a team like this work together is a special treat.

"Improvising with glass is a tricky thing, though. If you get too unscripted, things can go badly wrong, and glass is not a very forgiving material. So it's a delicate line . . . There are times when shit happens and the glass does something unexpected, and at those times it's really interesting to see how different gaffers and teams deal with the unexpected. Some ride with it and some break down and throw the piece away. Lino once said to me, 'It's not so much what you can make that determines how good a glassblower you are—it's what you can fix.'"

The manipulations that give rise to a finished piece can't be fully specified ahead of time. Rather, the piece is the frozen record of a team's coordinated finesse in responding to one another and to the glass. Having witnessed its making, I could only view the finished cane that Houk and the Demains produced that day as a sort of ecological specimen—a fossilized bit of joint attention.

What might follow from regarding it this way? Is there any consequence to my ecological metaphor? I believe it can inform our understanding of how competence arises, and help to clear away some misapprehensions about education that have deep roots in the West, and carry special peril in our current moment.

SCIENTIFIC DISCOVERY AS PERSONAL KNOWLEDGE

Michael Polanyi wrote, "An art which cannot be specified in detail cannot be transmitted by prescription, since no prescription for it exists. It can be passed on only by example from master to apprentice. This restricts the range of diffusion to that of personal contacts, and we find accordingly that craftsmanship tends to survive in closely circumscribed local traditions."[7]

Polanyi is here talking about craft knowledge, but he was seeking a larger epistemological point. Polanyi was one of the most prominent physical chemists of the middle of the twentieth century. In the second half of his life he took up philosophy in an effort to understand his own experience of scientific discovery. His elaboration of "tacit knowledge" entailed a criticism of the then-prevailing ideas of how science proceeds, tied to wider claims about the nature of reason. The logical positivists conceived reason to be rule-like, whereas according to Polanyi, a scientist relies on a lot of knowledge that can't be rendered explicit, and an inherent feature of this kind of knowledge is that it is "personal." He explained:

> The declared aim of modern science is to establish a strictly detached, objective knowledge. Any falling short of this ideal is accepted only as a temporary imperfection, which we must aim at eliminating. But suppose that tacit thought forms an indispensable part of all knowledge, then the ideal of eliminating all personal elements of knowledge would, in effect, aim at the destruction of all knowledge. The ideal of exact science would turn out to be

> fundamentally misleading and possibly a source of devastating fallacies.[8]

To understand what Polanyi has in mind when he speaks of "devastating fallacies," it helps to know something further about his life. A Hungarian, he was a refugee from the Soviet project to achieve rational planning of scientific research. The Communists' attempt to submit science to a five-year plan, in the service of social utility, spurred him to articulate a set of ideas that puts limits on any such project for the direction of research.

For Polanyi, scientific inquiry is above all a practice, best understood as a kind of craft. "I regard knowing as an active comprehension of the things known, an action that requires skill." He draws a parallel between science and craft that I take to be stronger than a mere analogy—rather, they are two expressions of the same mode of apprehending the world: by grappling with real things.

Writing after the war, he pointed out:

> While the articulate contents of science are successfully taught all over the world in hundreds of new universities, the unspecifiable art of scientific research has not yet penetrated to many of these. The regions of Europe in which the scientific method first originated 400 years ago are scientifically still more fruitful today [1958], in spite of their impoverishment, than several overseas areas where much more money is available for scientific research. Without the opportunity offered to young scientists to serve an apprenticeship in Europe, and without the migration of European scientists to the new countries, research centres overseas could hardly ever have made much headway.[9]

LIBERAL EDUCATION AS APPRENTICESHIP

It would be a gross misreading to take this as an expression of "Eurocentrism." Polanyi's point is that to have science, you must have scientists. Scientists are formed. They cannot be conjured wherever "money is available," or public purposes press (I take him to be referring to the United States).

In the time since Polanyi wrote, America has developed its own traditions. Entering graduate school in the natural sciences, at the end of your first year you join a lab and spend most of your waking hours for the next seven-odd years within its typically cinder-block walls. (A year of courtship is required so that researchers and new graduate students can form an assessment—of one another's character, not least.) Through long immersion in a particular field of practice and inquiry, you become a connoisseur of a certain class of intellectual problems. You adopt the language of your subfield, but also a shared, usually inarticulate sense of what sort of problems are worth investigating: what to take seriously. In the course of this apprenticeship you make the characteristic mistakes of a novice, and suffer their humiliations before your teachers (who include the more advanced graduate students). Conversely, you experience elation at those moments when you feel a growing mastery—you're becoming a journeyman. Through these experiences, theoretical thought and methodological tools get joined to a sense of personal involvement. It is not that you have simply swallowed a set of doctrines. Rather, the judgments of the discipline have become your own. Through such formation, you earn a certain independence.[10]

Such a culture of scientific apprenticeship has not yet developed in China or the Persian Gulf oil states. They have plenty of money and pressing public purposes, but these countries would seem to occupy a position similar to that of the United States in the middle of the twentieth century, when we depended on émigré scientists to help guide such endeavors as the Manhattan Project. Our scientists had access to the same handbooks of physical constants and the same textbooks and research journals and plenty of funding, but were not yet initiated into scientific inquiry as a mode of personal knowledge that is socially incubated, beginning with imitation. The Manhattan Project had a significant lasting effect by providing a setting for mentorship, which then reproduced itself in American universities. My own father was a beneficiary of these developments. After fighting in Europe he attended junior college on the GI Bill, transferred to U.C. Berkeley, and eventually joined the lab of Louis Alvarez, who supervised his doctoral work. This was the era of the bubble chamber, the advent of particle physics. My dad used to tell stories—some firsthand, some through the common lore of the lab—about the émigré figures who towered over physics in those days.

But the culture of scientific apprenticeship that developed in Europe, and then later in America, did so without warrant from the official self-understanding of modern science. As Polanyi writes, "To learn by example is to submit to authority. You follow your master because you trust his manner of doing things even when you cannot analyze and account in detail for its effectiveness."[11] This is intolerable if, like Descartes, you think that to be rational is to reject "example or custom" in order to "reform my own thoughts and to build upon a foundation which is completely my own." The paradox of the Cartesian project is that from a beginning point that is radically self-enclosed, one is supposed to proceed by an impersonal method, as this will secure objective knowledge—the kind that carries no taint of the knower himself. Polanyi turns this whole procedure on its head: through submission to authority, in the social context of the lab, one develops certain skills, the exercise of which constitutes a form of inquiry in which the element of personal involvement is ineliminable.

Let's dwell for a minute on the role that Polanyi assigns to trust: "You follow your master because you trust his manner of doing things." This suggests there is a *moral* relation between teacher and student that is at the heart of the educational process. Of course, the student must trust that the master is competent. But he also must trust that his intention is not manipulative. It is the absence of just this trust that we found at the origins of Enlightenment epistemology in the previous chapter: a thorough rejection of the testimony and example of others. This rejection begins as a project for liberation—from manipulation by kings and priests—and blossoms into an ideal of epistemic self-responsibility. But the original ethic of suspicion leaves a trace throughout. This stance of suspicion amounts to a kind of honor ethic, or epistemic machismo. To be subject to the sort of authority that asserts itself through a claim to knowledge is to risk being duped, and this is offensive not merely to one's freedom but to one's pride.

If Polanyi is right about how scientists are formed, then the actual practice of science proceeds *in spite of* its foundational Enlightenment doctrines: it requires trust. The idea that there is a method of scientific discovery, one that can be transmitted by mere prescription rather than by personal example, harmonizes with our political psychology, and this surely contributes to its appeal. The conceit latent in the term "method" is that one merely has to follow a procedure and, *voilà*, here comes the discovery. No long immersion in a particular field of practice and inquiry is needed; no habituation to its peculiar aesthetic pleasures; no joining of affect to judgment. Just follow the rules. The idea of method promises to democratize inquiry by locating it in a generic self (one of Kant's "rational beings") that need not have any prerequisite experiences: a self that is not situated.

Polanyi saw an alliance between this misapprehension of scientific practice and efforts to direct research according to some societal goal, for example a Soviet five-year plan of scheduled technological breakthroughs. After many forms of tradition and local knowledge were deliberately destroyed in China during the Cultural Revolution (a spasm of hyperenlightenment), one can imagine the frustration of the current Chinese regime as it now pours billions into its universities, hoping for discovery and innovation but instead producing rule followers.

The question arises how our own intellectual traditions, both scientific and humanistic, will be affected by the current transformation of the American university along the lines of a business enterprise. We are told that there are exciting efficiencies to be realized by replacing face-to-face instruction with Massive Open Online Courses (MOOCs). However appropriate—even ideal—they may be for instruction in some narrow technical matters (I am a big user of YouTube instructional videos on topics like computer-aided design, and how to build electronic fuel injection systems), in the arts and sciences we should take notice that MOOCs divorce the articulate content of a field from personal interaction with a teacher who has made it his vocation to live with the field's questions. There is, then, a certain harmony between these institutional developments and our deep supposition that the ideal of perfect "clarity"—of precise formalization—is both possible and desirable and that, if realized, it would make any field transmissible by impersonal means. But let us heed Polanyi's warning that "the ideal of eliminating all personal elements of knowledge would, in effect, aim at the destruction of all knowledge."

Polanyi's argument about the role of unspecifiable, tacit knowledge in expertise; his elaboration of personal commitment as the core of intellectual inquiry, understood as a craft skill; his demonstration that scientific competence is transmitted through apprenticeship to authoritative teachers—from all of this, "it follows that an art which has fallen into disuse for the period of a generation is altogether lost." He goes on:

> There are hundreds of examples of this, to which the process of mechanization is continually adding new ones. These losses are usually irretrievable. It is pathetic to watch the endless efforts—equipped with microscopy and chemistry, with mathematics and electronics—to reproduce a single violin of the kind the half-literate Stradivari turned out as a matter of routine more than 200 years ago.[12]

Cultural Revolutions aren't imposed only by totalitarian regimes. We call ours "the creative destruction of capitalism," and shower venture capital on "disruptive technologies," especially ones that promise to mechanize human interaction. It hardly needs to be said that the results are both positive and negative. But it does need to be said that in the university, the survival of our traditions of intellectual apprenticeship should not be taken for granted. They will not be well equipped to defend themselves against the Maoist MBAs if they are not aware of themselves *as* traditions, but remain wed to a conception of knowledge as something that is transmissible to atomized individuals, without loss. On that conception, there is no clear reason why one ought to be creeped out by the idea of a central repository of knowledge, to which we are all connected in a Massive Online intellectual life.

I am not much inclined to defend undergraduate education in its current form, and have expressed a fairly jaundiced view of its role in society elsewhere.[13] But one of the things you learn in studying the history of politics is that power is consolidated by eliminating intermediate structures of authority, often under the banner of liberation from those authorities. In his book *The Ancien Régime and the Revolution*, Tocqueville gives an account of this process in the case of France in the century preceding the Revolution. He shows that the idea of "absolute sovereignty" was not an ancient concept, but an invention of the eighteenth century that was made possible by the monarch's weakening of the "independent orders" of society—self-governing bodies such as professional guilds and universities. The revolutionists inherited this (fairly recent) centralization of power from the monarchy, and now defended such centralization as the guarantor of liberty against all intermediate forms of social authority (the kind exercised by independent associations). The ideal of total liberty required total centralization of power, now in the name of the people. Today it is the vanguardist disrupters at Google who promise to deliver us from parochialism. If Polanyi is right about how scientists and other thinkers are formed, then to weaken the local authority of teachers and traditions that embody "personal knowledge" is a bad idea, on both epistemic and political grounds.

It is part of our Enlightenment heritage that we are taught to take an intransigent stance against the authority of other people. In the budding romance between Silicon Valley and our universities, there is an exciting prospect that "the scent of people might be removed altogether" (as Jaron Lanier said in another context). If you can't smell it and you can't touch it, whatever authority is acting must be that of reason itself! Quite apart from the business appeal of MOOCs for universities (payroll is a lamentable thing), mechanizing instruction is appealing also because it fits with our ideal of epistemic self-responsibility.

As we will see shortly, this aspiration to self-responsibility is at odds with some elementary facts about human beings, in particular the role that other people play for us in conditioning the way we grasp the world.

7

ENCOUNTERING THINGS WITH OTHER PEOPLE

We have already considered how embodiment plays a fundamental role in perception. Things show up for us in daily life, not as they would to a disinterested spectator, but as things that we "have to do with" in some way. It follows that when we become skilled in some particular domain, we begin to see and feel things we otherwise wouldn't see or feel. The world acquires new "affordances" that guide us in what amounts to a new ecological niche that we have begun to inhabit. This new ecological niche is a new space for action. The most basic point of Part I was that our cognitive capacities are those of a being who develops from infancy not as a pure observer of the world, but as one who acts in it.

In Part II we are focusing on another basic fact that contributes to our "situatedness": *The world that we act in is one that is inhabited by other people.* As infants, we find ourselves thrown into the world midstream; it is already saturated with sediments of meaning that have been building up in the social world since long before we came along. It is not simply that other people are among the objects we perceive. Rather, others set up shop in our consciousness in ways that condition how we perceive and use everything. One way to begin approaching this idea is by revisiting the concept of affordances, and extending it from our physical environment to our cultural environment.

FROM AFFORDANCES TO EQUIPMENT

Consider a pair of slender sticks about eight inches long. When I see them on the table in an Asian restaurant, I reach for them in order to bring some noodles to my mouth. But in the context of a percussion class for toddlers, they would likely show up as toddler-sized drumsticks of the lightweight sort used for timbales. In these different contexts, the sticks are part of different "equipmental wholes," to borrow a concept from Martin Heidegger. The idea is that equipment always refers to other equipment, and to a set of social practices that more or less coheres. Only within those social practices does the individual object show up as useful.[1] Thus chopsticks are part of a practice of dining that includes, for example, the use of bowls rather than plates, and the preparation of sticky rice rather than, say, loose peas.[2] Chasing peas around a plate with chopsticks, or trying to eat a porterhouse steak with them at a Western-style hotel restaurant in Beijing, one can't help but exclaim, "What the hell? These things are *useless*."

Chopsticks belong to a different equipmental whole than forks and knives. Their usefulness is not simply a function of fit between a person who acts and a single object taken in isolation; nor is this fit determined by a purpose that is simply the actor's own. Rather, in using things like chopsticks, or fork and knife, we involve ourselves in *norms*: it is just understood that one does things a certain way. These norms are for the most part inarticulate; they are tacit in social practices and in the equipment we use. This is one way in which other people condition the way the world presents itself to us, even when we do not interact with them.

Take a moment to look at the walls in whatever room you are in. I am at a library, and the walls are a slightly yellowish beige. If asked their color, that is what I would say, and not give it another thought. If I look more closely and get analytical about it, I notice that in the particular lighting conditions in this room at a certain time of day—and today is a bright day in April rather than a hazy day in August—different parts of the wall are "really" (that is, empirically) different shades of beige, some darker and some lighter, some with a glare from the window, some illuminated by an overhead fluorescent lamp, some closer to the warmer-colored reading lamp, and others in shadow.

But this isn't to clarify my original perception of the wall (as uniformly beige) so much as it is to substitute a new perception for the original one. This new perception is the one that empiricism likes to talk about. Maurice Merleau-Ponty writes that empiricism "is not concerned with what we see, but with what we ought to see, according to the retinal image."[3] But before getting analytical about my visual experience, I simply saw a beige wall. This suggests that whatever it was that was determining my perception in that original case, it wasn't *simply* stimuli as understood by the empiricist.

To begin with, my previous experience enters into my current experience.[4] I have experience with how the appearance of things varies in reflected light from various sources, one of which (the sun) moves in certain characteristic ways according to the season and the time of day, and is partially scattered and occluded by clouds of various descriptions. A baby hasn't yet learned these things, and one suspects it perceives something more riotous than a wall that is simply beige. But as an experienced perceiver, I *extract invariants from the flux of stimuli*, to use Gibson's formula, and it is the invariants that I perceive, unless I make an effort to do otherwise (as the artist must—she applies different colors to the canvas to represent that uniform wall).

Further, my experience of the world includes experience of other people—the world I inhabit is a shared world, and this is a very basic feature of it. If one takes this into account when considering the problem of "color constancy," the limitations of empiricism's view of perception become clear. I am not a baby or a blank slate. As an acculturated member of society, I happen to know how painters proceed when painting a wall. They don't carefully draw geometric shapes with slightly different shades here and there; they show up with five-gallon buckets and knock it out. This is not something I consciously think about when I am sitting in a beige room, going about my business. But I have this stock of social knowledge, and it seems to condition my immediate perception of the wall as uniformly beige. The point of calling this perception "immediate" is to claim that it does not depend on a process of interpretation that is laid upon, or comes after, elementary perceptions.[5]

The uniformity of the wall's color is a *social* fact, and what I perceive, in everyday life, seems to be such social facts, rather than the facts of optics. The facts of optics are not being challenged here, but their role in understanding human perception is limited. Though they always have some phenomenological significance, they are dispositive only in special cases, when we task ourselves with perceiving in a special way. To perceive the wall as variously colored, I have to suspend my normal socially informed mode of perception.

This is what an artist does. She must defamiliarize herself with her everyday perceptions, which depend on—are conditioned by—her past experiences, including the experience of inhabiting a world that is thoroughly conventional. She has to try to perceive as a baby does, or as the empiricist supposes we all do, but this is a subtle and extraordinary accomplishment. There is nothing infantile about good art, but it does show us the world as viewed by a consciousness that has, for a spell, liberated itself from conventionality.

The critical point against empiricism, then, is that we are social and biographical beings, not digital cameras or recorders. As I sit here writing, in a library, I hear a noise coming from somewhere above and behind me. Though I could do so, it would be a specialized sort of exercise, alien to everyday life, for me to describe it in purely naturalistic terms—as, say, a frequency distribution of compression waves arriving at my ear. If you ask me what I hear, I'll say "the ventilation system." I live in a society in which there are such systems in buildings, and this fact enters prereflectively into my immediate apprehension of the sound as what it is; I don't have to add a layer of interpretation to sensual data that is somehow experientially prior to HVAC systems. This sound, which in another context I might hear as the wind in the trees, shows up for me now against the background of a set of social practices and norms that govern the construction and daily operation of library facilities such as the one I am sitting in. All of this taken-for-granted social knowledge enters into my perception—I hear the HVAC system.

This way of naming my experience is the most phenomenologically rigorous. Again, I could describe the sound in naturalistic terms, but in doing so I would rely on a set of theoretical posits. Not least among these is the assumption of a naive, unacculturated, strictly individualistic hearer of sounds. This mythical creature is familiar to us all—it is the human person as conceived in the tradition of epistemic individualism that extends from Descartes's *Meditations* through eighteenth-century empiricism and on to contemporary cognitive science.

We live in a world that has already been named by our predecessors, and was saturated with meaning before we arrived. We find ourselves "thrown" into this world midstream, and for the most part we take over from others the meanings that things already have. How are we initiated into these meanings? This question leads us into fascinating issues in developmental psychology.

JOINT ATTENTION

In the first weeks of life, a human infant and its caregiver attend to one another intensively, staring into each other's eyes, smiling at each other, and copying each other's gestures.[6] Around the age of six months, the baby begins to direct its attention beyond this pairing and attend to the same object as its caregiver by following her gaze. Shortly thereafter, the baby begins "gaze-checking" with its caregiver if its initial gaze-following doesn't lead to some object that seems worth paying attention to.

The capacity for "joint attention," emerging sometime around the age of twelve months, entails something further. At this stage, the child has "an ability and willingness to enter into episodes in which there is a third object that mother and child are attending to jointly, with mutual understanding of the fact that the attention is shared," as Christopher Mole puts it.[7] This stage seems to coincide with the child's dawning awareness that its caregiver's utterances aren't just sounds; they refer to things in the world. Joint attention is thus intimately bound up with the capacity for communication, which requires not only awareness of the existence of other minds, but mutual awareness of a joint field of reference: the shared world.

At roughly the same developmental stage, occurring around twelve months, the child's pointing gestures take on an intentional character.[8] Two varieties of pointing emerge: imperative pointing, in which the child makes a request for some object, and declarative pointing, in which the child tries to get an adult to engage with its own attention to an object. "Attend to my attending" seems to be the demand conveyed by declarative pointing.

Here, then, is a developmental account of the social reflexivity that underlies our ability to communicate about the world, which seems to be distinctive to human beings. Chimpanzees, for example, do not exhibit declarative pointings.[9] Jane Heal writes that "words are, on this conception, an immensely delicate and useful way of pointing. Pointing itself is an elaborated way of focusing shared gaze. And what in turn grounds the whole enterprise is the sense of living together with another . . ."[10]

As Heal points out, it is in episodes of cooperation in some practical activity, such as playing with blocks together, that joint attention is focused and becomes an occasion for communication, such as "Oh look, the blue one has fallen over!"

The fact that we live together in a shared world, and do things together, is fundamental to the kind of beings we are. As Axel Seemann writes, "The recent surge of interest in joint attention attests to a shift away from a solipsistic conception of mind and toward a view of mental phenomena as inherently social."[11]

These insights from developmental psychology tend to deflate certain problems that have occupied people in philosophy of mind since the time of Descartes. One question that loses its force in this developmental perspective is that of how common knowledge is constituted, or how it is possible, as it must be for us to cooperate. A prominent approach to this puzzle is to suggest that if you and I are looking at a tree (the same tree), the way we can know it as being a joint object of perception for us is as follows: I see the tree, I *believe* that you see the tree, I further believe that you believe that I see the tree, and that you ascribe a similar train of believings to me.

This is implausible enough as an account of what we are doing (what, subconsciously?) when we blithely proceed, as we do, through our shared world in which we are able to communicate about the tree. Maybe each of us takes one end of a two-handled saw and cuts the thing down. Arguably we do this without forming explicit beliefs, or having to engage in mind reading.[12] Where is the problem? More particularly, this iterated-ascription-of-beliefs account runs up against a robust finding in developmental psychology, namely that children are unable to grasp the concept of *others' beliefs* until about the age of four, long after they have developed the capacity for a reflexively mutual awareness of the world, as revealed by declarative pointing.[13]

It is in social interaction that our mental capacities develop, to begin with, and this fact seems both to secure the availability of our minds to one another and to order the way in which we apprehend the world altogether.

This is a theoretical point, but it has very real consequences. Insofar as empiricism and other forms of epistemic individualism misdescribe our experience, they tend to cause mischief. The supposed infallibility of the "eyewitness," for example, is an entrenched assumption, but psychology has in recent decades become sophisticated about the limitations of this kind of testimony.[14] Awareness of these limitations hasn't much penetrated the legal system, however, as it is at odds with a judicial culture that often seems to value sheer volume of convictions over justice, and therefore favors rules of evidence predicated on simplistic views of cognition. Given the uses to which they get put, bad epistemologies are not culturally innocent.

The phenomenologist Alfred Schutz pointed out that our sensual memories, such as that of the eyewitness, fade quite quickly, but they also get idealized according to social norms, and in doing so they actually become more vivid (even if false); they become something that one can hold on to. Language plays a decisive role in this process: we articulate our experiences. We do so in the particular language we are born into, making use of the prevailing stock of ready-made phrases that currently circulate. In doing this we subject them to "typifying schemata of experience." These typifications both idealize and socialize our originally private, sensual experience.

This may help to explain how social stereotypes, which we articulate in speech, infect eyewitness testimony. Or consider the fact, now widely known, that the web of norms and expectations that get conveyed in conversations with social workers and other therapeutic professionals can implant false memories in people—most wrenchingly, about child abuse. Through social typifications in language, our memories get bent toward whatever is allowed or encouraged by authoritative voices or by the larger swirl of democratic opinion.

WHAT ABOUT INDIVIDUALITY?

In light of all this, how are we to understand individuality, taken both as a fact (we are all different) and as an ideal that we cherish? Given all the ways that others set up shop in our consciousness, it seems hard to mark out "conformity" as some kind of ethical failure.

The problem is a deep one. I have argued that internal, private mental experience is *not* what is initially and most certainly given to us. Things in the world show up for us in already established meaning contexts that we were initiated into as young children, such as the perception of a uniformly colored wall or an HVAC system. Tools are usually not the implements of an isolated person acting on the world; their physical affordances refer to a whole set of related equipment, and thereby to social norms and practices (as in the case of chopsticks). Through the conventionality of the language we use to describe them, our initially sensual memories get bent toward social norms. The deep point is that our private experiences are founded on—would not be intelligible without—the prior disclosure of a shared world. This is the world we encounter first, as babies locked in joint attention with a caregiver.

It follows that our experiences are not simply "our own." This is a bit alarming, perhaps. One response would be to double down on epistemic individualism, and emulate Descartes in his efforts to achieve independence of mind by excluding the testimony of others. But this is unrealistic, for all the reasons we have explored.[15] My hope is that developing an alternative picture of our mental lives, one that does justice to our nature as social beings, can help illuminate the grounds on which individuality really *is* possible—not solipsistically but sociably, in practices that bring us into cooperation with others. Individuality is something that needs to be achieved, and in this endeavor other people are indispensable to our efforts.

8

ACHIEVING INDIVIDUALITY

There is a certain modern type who makes a hobby of his or her "inner work." Their talk tends to be a mash-up of therapeutic and New Age idioms. If by some device such a person has succeeded in putting you under an obligation to listen to her talk about herself, you may find yourself playing a more important role for her than you would prefer.

She tries to get you to endorse what strikes you as a fairly elevated view of herself, one that she has evidently spent a good bit of energy working up. This is a delicate sort of conversation to manage. You try to simply go along with it, amiably. But then she accuses you of not taking her seriously, not really *engaging*. This makes her angry. On the one hand, she insists that she has privileged access to the truth about herself. Through a process of introspection, she discovers in herself motives and character traits that, precisely because they are discovered by introspection, are not contestable by others. But on the other hand, she needs this truth to be validated. By you.

The German philosopher Georg Wilhelm Friedrich Hegel was well aware of the inner-outer motif and the function it serves in carving out a realm that is safe for self-delusion. He anticipated Bruce Springsteen, who is reported to have said, "Self-knowledge is a kind of funny thing because the less of it you have, the more you think you have."

For Hegel, one knows oneself by one's deeds. And deeds are inherently social—their meaning depends very much on how others receive them. The problem of self-knowledge is in large part the problem of how we can make ourselves intelligible to others through our actions, and from them receive back a reflected view of ourselves.

For Hegel, there is no self to be known that exists prior to, or at a "deeper level" than, the self that is in the world. This implies that individuality, too, is something that we achieve only in and through our dealings with others.

Right away, there is cause for concern in this idea that the self is constituted by its actions. What if you find yourself playing a role prescribed by your social situation, and what is required of you is some act that you're not really able to stand behind? Suppose you are a citizen of North Korea and Kim Jong Il has just died. Do you join in the wailing and hysterics? Yes you do. Closer to home, suppose you are attending a child's birthday party. As the voracious tyrant rips open yet another gift, you find yourself cooing "Good job!" along with all the other mommies and daddies. What has to be the case for you to be able to identify with your actions as really being your own? Surely we would want to say that your true self is the one that is revealed, and perhaps emerges in the first place, through a particular *kind* of action—the kind that isn't alienated.

Some will say that sincerity is the key element here; whether or not an act is a true expression of the self is determined entirely by the inner psychological state of the agent. But consider the case of someone who yells "Fire!" in a crowded theater and a few people get trampled to death (this example is from a book on Hegel by Robert Pippin). This person regards himself quite sincerely as a well-meaning prankster—people ought to lighten up! But when he later says, "I was just joking," his claim to have been misunderstood is not just self-serving, it is also self-deluded. One must know something about how the world works, about the norms of speech and action that prevail in a given society, and understand one's own actions in light of these. The agent himself cannot unilaterally say "what was done" and thereby trump the meaning that his deed has for others.

Pippin puts Hegel's point sharply when he writes, "You have not executed an intention successfully unless others attribute to you the deed and intention you attribute to yourself."[1] One can think of counterexamples to this formula—a successful deception, for instance. But it serves well as a corrective to the cult of sincerity, which perhaps amounts to this: the idea that you yourself can be the source of the norms by which you justify yourself. This idea seems to be the late modern understanding of autonomy, in a nutshell.[2]

Hegel says we need other people as a check on our own self-understanding. Our deeds bring us out into the light of day, and the way others receive them helps us triangulate on a true assessment of ourselves. This makes me think of economics. In economics, when we talk about the *value* of something, we are referring to an assignation of worth that has to be shared in order to be determinate. That is what it means for something to have a price. I want to bring out the affinities between economics and Hegel's critique of sincerity by applying his logic of self-knowledge to a certain kind of economic exchange: getting paid for work you have done. If it is by our deeds that we know ourselves, then to be paid for your deeds—not the alienated kind, but the kind you are able to stand behind and claim as your own—would seem to be a good Hegelian recipe for discovering one's worth. There are good reasons to have reservations about such a recipe, and we will consider them. But I think there is some psychic reality to be explored here. I want to consider how the simple acts of demanding and receiving payment for work may carry the possibility of self-knowledge, and in some cases may be the kind of encounter in which individuality is not merely revealed, but forged.

JUSTIFICATION THROUGH CONFRONTATION

Consider the case of a motorcycle mechanic. In handing a labor bill to a customer, I make a claim for the value of what I have done, and put it to him in the most direct way possible. I have to steel myself for this moment; it feels like a confrontation. The point of having a posted labor rate, and hours billed in tenths on the service ticket, is to create the impression of calculation, and to appeal to the authority of an institution with established rules. But this is a thin and fragile pretense observed by me and my customer (it is hard to pose as an institution when you run a one-man operation), and in fact the bill I present is never a straightforward account of hours worked. It always involves a reflection in which I try to put myself in the shoes of the other and imagine what he might find reasonable.

This lack of straightforwardness in valuing the work is due to the fact that the work is subject to chance and mishap, as well as many diagnostic obscurities. Like medicine, it is what Aristotle calls a "stochastic" art. Especially when working on older bikes, in trying to solve one problem, I may create another. How should I bill for work done to solve a problem of my own making? Should I attribute this new problem to chance, or to a culpable lack of foresight on my part? This question has to be answered when I write the service ticket, and in doing so I find that I compose little justificatory narratives.

When a customer comes to pick up his bike, I usually go over the work with him in detail, and I often find myself delaying the presentation of the dollar amount, because I fear that my valuation isn't justified. But all my fretting about the bill has to get condensed into a definite assertion on my own behalf. Whatever conversation may ensue, in the end the work achieves a valuation that is determinate: a certain amount of money changes hands. As he loads his bike onto the back of a pickup truck, I want to feel that the customer feels he has gotten a square deal; I want to come away feeling justified in the claim I made for the value of what I did.

Here, in a microeconomic exchange, lies the kernel of ethics altogether, perhaps. In presenting the labor bill, I am owning my actions. I am standing behind them retrospectively. And this requires making my actions intelligible to the customer. The Hegelian suggestion seems true to me—namely, that it is in the confrontation between the self and the world beyond one's head that one acquires a sharpened picture of each, under the sign of responsibility.

As Pippin explains, what distinguishes human acts from mere events, and from animal doings, is that we are concerned with justification. Our deeds don't simply enact our desires. Rather, in acting we make a tacit normative claim for ourselves—for the legitimacy of the act, and indeed the worthiness of its end. Crucially, Hegel suggests that this normative moment arises only in a certain kind of encounter with another person—someone who addresses me, or issues a sort of summons to tell him "where I am coming from." More strongly: the question of justification arises only if I am *challenged* by another, or anticipate being challenged by another, who doesn't merely stand in my way as an impediment to my doing what I want to do, but rejects the *validity* of my claim to be acting with justification. To rise to this challenge means I have to evaluate my own actions—are they something I really want to assert the worth of? Stepping back and considering in this way is something I have to do if I am going to own my deeds; if I am going to stand behind them and identify with them as my own, and not regard them as mere movements that my body has made. It is this evaluative stance toward ourselves that distinguishes human beings.

Work, then, is a mode of acting in the world that carries the possibility of justification through pay. When the claim I make for the value of what I have done prevails in a meeting with another free agent and I succeed in getting paid, I take this as a validation of my own take on my doings.[3] The absence of such experiences may help us to understand why the long-term unemployed often suffer self-doubt, as do the idle adult children of wealthy parents.

THE PROBLEM OF THE DISSIDENT

But consider also that this validation through pay is a function of the prevailing political economy. As Talbot Brewer put it to me in conversation, the politico-economic regime may reflect back a distorted view of oneself. It may confer an inflated salary and corollary self-regard on some professions while placing a slight value on others, being indifferent or oblivious to the excellences these latter demand of their practitioners. Because we are social creatures and refer ourselves to others for justification, such obliviousness in the larger society may infect a man's own experience in such a way as to make it illegible to himself.

Every regime has such blind spots and exaggerated valuations with regard to the range of human possibilities. They have a political character to them, shaping souls and forming the young in the image of the regime. Imagine a high-achieving university student who understands that he is supposed to want to be an investment banker but is left cold by the picture of his future that comes into view when he imagines such a life. He would really rather be building houses, having gotten a taste of that life while working in construction one summer. But he finds it difficult to articulate what he finds valuable about this activity and to justify it as a choice of livelihood in the terms prevailing in the public discourse, or given the expectations of his social milieu. So he brackets as best he can these unsanctioned intimations of what a good life for himself would look like, and with the help of a little medication they wither, like a limb that has been tied off to prevent an infection from spreading.

On the Hegelian position elaborated by Pippin, there would seem to be little room for dissent from the mainstream. To be an agent in the full-fledged sense is to be well adjusted to social norms, because these provide the only possible justificatory framework for one's deeds. Absent the public framework, one is at sea without a compass or keel, listing badly toward fantasy. The fate especially to be avoided is that of Don Quixote, who takes himself to be a knight engaged in acts of chivalry but inhabits a social world in which such roles and deeds are not possible. That is, they are not recognized, not intelligible to others.

This is a deeply conformist line of thought. It leaves little space for beautiful folly, or for the world-making activity of the artist or eccentric. Yet Hegel's central insight into the social character of genuine agency (and corollary worry about solipsism and self-delusion) seems to me on target.

The question I would like to pose, then, is this: To whom does one look for a check on one's own subjective take? To "the public," or to the *competent* within some concrete community of practice?

There are many such communities, corresponding to diverse niche ecologies of human excellence, while the public is an undifferentiated blob.

Bringing our focus down to a smaller scale in this way won't secure space for the genuine maverick, but I hope it will help to articulate the grounds on which people who have come together around some practice—one that has formed them in important ways and perhaps leaves them feeling untimely, or out of joint with the surrounding society—might carve out normative niches for themselves, resisting the imputation of insanity and defending themselves against the functionaries of psychological adjustment. The practices I have in mind, as being especially countercultural and therefore in need of defense, are philosophy and craftsmanship.

ON THE "WHO" AND THE "WHAT" OF JUSTIFYING NORMS

For Hegel, the "who" with reference to whom one justifies one's actions are those who are similarly habituated within some particular form of ethical life—a cultural jig that has developed over time and offers a meaningful frame for one's activities. In such a world, deeds have a revelatory power. They speak for themselves, and this is because they are addressed to, or potentially taken up by, others who inhabit the same culture, within which deeds have somewhat fixed meanings. Sacrificing a lamb is intelligible (*as* a sacrifice rather than, say, as performance art) only if a whole set of cultural enabling conditions are in place.

But this means that in times of cultural flux and uncertainty, when it is not clear what "our rules" are, there is a basic difficulty for individual agency understood socially. One is thrown back into oneself, with little reference beyond the movements of one's own will and solitary judgment.

Under such conditions, the material practices of making things and fixing things take on special significance. Their meaning does not depend on fragile cultural conditions and shifting articulations. If we are dealing with concrete stuff together, our actions are likely to achieve the mutual intelligibility that is required for genuine agency. Matt Feeney said it well: "The nature that is providing for and vexing me is the same nature that is providing for and vexing everyone."[4]

Note that this is true of nature only in a trivial sense if the "everyone" is not further specified—in that case we are limited to such banalities as "What goes up must come down." Nature provides a meaningful ground for mutual intelligibility only if you get more specific, and this is what happens within a community of skilled practice. Competence rests on an apprehension of real features of the world, as refracted through some set of human needs/desires and corresponding technologies. These features may be easy to grasp, as when a master plumber shows his apprentice that he has to vent a drainpipe in a certain way so that sewage gases don't seep up through the toilet and make a house stink. Or it may be something requiring subtle discernment, as when a better motorcyclist than I explains, from a rider's point of view, why it would be good to decrease the damping in the front suspension of his motorcycle. There is a progressive character to these apprehensions—something about the world is coming into clearer view, and your own judgments are becoming truer. Or rather, you are becoming more discerning, seeing things about which you had no judgments previously. Getting outside your head in this way, you have the experience of joining a world that is independent of your self, usually with the help of another person who is further along. The process is aided if (as in the case of suspension tuning) the pertinent features of the physical system you and he are grappling with are apprehended through sensorimotor engagement, as then the desired qualities of the system can become an object of joint attention, and hence of communication.

This kind of education is driven forward by the visceral concern for excellence that comes with being initiated into the practice, and a corollary contempt for the shoddy (for example, a toilet that stinks or a bike that wallows through turns). In the course of becoming skilled, feeling is joined to judgment, and our perception becomes evaluative: the ends served by the practice illuminate our activity, casting deep shadows wherever our performance fails to live up to them. To the extent these ends are simply functional they are graspable by anyone; they are "public" in that sense and provide the grounds for getting paid. But meeting the standards of function doesn't exhaust what the practitioner cares about in doing what he does. Because he strives for excellence, there is room for a kind of freedom and individuality in the practitioner, even as he conforms to the public, merely functional norms of the practice.

A carpenter, for example, answers to his level, his square, and his plumb bob: standards that have universal validity. Yet it is up to him how these minimal standards will be met in the graceful arc of a stairway's handrail. The discriminations made by practitioners of an art respond to subtleties that may not be visible to the bystander. Only a fellow journeyman is entitled to say, "Nicely done." In doing the job nicely, the tradesman puts his own stamp on it. His individuality is thus expressed in an activity that, in answering to a shared world, connects him to others—in particular, to other practitioners of his art, who are competent to recognize the peculiar excellence of his work.

Earlier, when I considered the example of the mechanic, I suggested that confronting a customer with a labor bill is a moment when you have to step up and justify your deeds to another, and that doing this confers on your work the status of nonalienated action. That is, it is action you are able to assert the value of to another, and take responsibility for. If you succeed in getting paid, the value you asserted is validated not just by the customer, but by all who hover in the background of the transaction: the entire market for similar services.

But then I had some second thoughts, when we considered the kind of distortions that are typical of the marketplace, where some activities receive inflated valuations and others are slighted. The values assigned by the marketplace are an unreliable proxy for human excellence. Even in its ideal form, the "free market" can offer only an abstract sort of valuation, since it is predicated on fungibility. The great heterogeneity of goods and services is treated as fundamentally equivalent; each can be represented by a price—a point on a shared scale that stretches over a single dimension. The market is leveling, whereas our evaluative activity as moral agents is sensitive to differences of kind.

Therefore the exchange between the mechanic and his customer can go only so far in satisfying the mechanic's need for validation of his *peculiar* excellence, as a mechanic, for the simple reason that the customer isn't competent to recognize its finer points. What the mechanic wants—what we all want—is recognition. But that is something you can get only from your peers; from people whose vision has been sharpened and sensitized to the relevant considerations through a process of initiation.

What we want, when we want recognition, is to be recognized as an individual. This seems to be possible only in the context of genuine connection to others, with whom one is locked into some web of norms—some cultural jig—that is binding, yet also rich enough to admit of individual interpretation. Skilled practices fit this description, and for that reason have special significance in our efforts to win recognition as individuals. Our efforts on that front get confused and misdirected when we live under a public doctrine of individualism that systematically dismantles shared frames of meaning. The reason we need such frames is that only within them can we differentiate ourselves as not merely different, but excellent. Without that vertical dimension, we get the sameness of mass solipsism rather than true individuality.

The de-skilling of everyday life, which is a function of our economy, thus has implications that reach far beyond the economy. It is integral to a larger set of developments that continue to reshape the kinds of selves we become, and the set of human possibilities that remains open to us.

9

THE CULTURE OF PERFORMANCE

In *The Weariness of the Self,* Alain Ehrenberg offers a cultural history of depression. He writes:

> Depression began its ascent when the disciplinary model for behaviors, the rules of authority and observance of taboos that gave social classes as well as both sexes a specific destiny, broke against norms that invited us to undertake personal initiative by enjoining us to be ourselves . . . Depression presents itself as an illness of responsibility in which the dominant feeling is that of failure. The depressed individual is unable to measure up; he is tired of having to become himself.[1]

In the 1960s, personal liberation—from the authority of parents, teachers, bourgeois laws, the uterus, the draft, the bra—happened to coincide with a period of upward mobility in a booming economy. These developments seemed, for a moment, to herald the arrival of the strong one prophesied by Friedrich Nietzsche. Ehrenberg quotes from *The Genealogy of Morals*: "The proud knowledge of the extraordinary privilege of responsibility, the consciousness of this rare freedom, of this power over himself and over fate, has sunk right down to his innermost depths, and has become an instinct, a dominating instinct." For some decades now, this sovereign individual has been the stock character described in commencement speeches. It is the background picture of the self that informs daytime talk shows and advice columns. It is what a high school guidance counselor falls back on when his blood sugar is low.

The sovereign individual has become our norm but, as Ehrenberg says, "instead of possessing the strength of the masters, she turns out to be fragile, . . . weary of her sovereignty and full of complaints."[2]

THE CULTURE OF PERFORMANCE

Our weariness is understandable. With radical responsibility comes a new emphasis on personal initiative, and a corollary "culture of performance" in which you have to constantly marshal your internal resources to be successful, as Ehrenberg says. This is reflected in, for example, the heightened competition of the middle-class educational trajectory. Significant social sorting is understood to be operating at every stage, from preschool to the GREs. With our presumption of meritocracy—that is, of a fair and frictionless mobility, a system without any systemic rigidities that would block our way—failure carries a deeper stigma than it would if we had a more realistic view of our society.

If there are no external constraints, what you make of yourself depends on your gumption and mental capacities. Are you a high-performance person? In a culture of performance, the individual reads the status and value of her soul in her worldly accomplishments. Like the Calvinist, she looks to her success in order to know: Am I one of the elect or am I damned? With radical responsibility comes the specter of inadequacy.

In Calvin's time, one might have had a hereditary occupation. And as recently as the 1970s, it was possible to compose a working life centered around the steady accumulation of experience, and be valued in the workplace for that experience; for what you have become. But, as the sociologist Richard Sennett has shown in his studies of contemporary work, it has become difficult to experience the repose of any such settled identity.[3] The ideal of being experienced has given way to the ideal of being flexible. What is demanded is an all-purpose intelligence, the kind one is certified to have by admission to an elite university, not anything in particular that you might have learned along the way. You have to be ready to reinvent yourself at any time, like a good democratic *Übermensch*. And while in Calvin's time the threat of damnation might have been dismissed by some as a mere superstition, with our winner-take-all economy the risk of damnation has acquired real teeth. There is a real chance that you may get stuck at the bottom.

MOBILITY AND THE DEMOCRATIC SOCIAL CONDITION

When Tocqueville came to America in the 1830s, he was struck by our "democratic social condition," in which "new families are constantly springing up, others are constantly falling away, and all that remain change their condition." Social mobility represents a *possibility*, a powerful *idea* of equality that carries psychic force even if you find yourself (for now) near the bottom.

In our daily intercourse there is a fairly easy mixing of people of different fortunes, without set rituals of deference and condescension. A certain democratic amiability is expected of all. "How's it going?"—thus does the valet address the guy whose Ferrari he is about to park. For his part, the Ferrari owner feels flattered rather than offended by the familiarity. He prides himself on being a regular guy, and is put at ease by the exchange. He should leave a fat tip; the valet has done him a psychic favor. But wait—a big tip would draw attention to the economic inequality, and thereby undermine the whole exchange, which is a mutual performance of social equality. It's complicated in America. (To get by in the service industry here, you have to learn to finesse this stuff to your advantage. You have to play on the Ferrari owner's democratic virtue before handing him back the keys.)

Recent opinion surveys indicate that Americans still "have a greater faith in their country being a meritocracy than citizens of nearly every other country on earth," according to *The Huffington Post.* Yet recent measures of equality of opportunity and social mobility from one generation to the next place us dead last among the advanced nations studied, which include the United States, the UK, France, Germany, Sweden, Italy, Australia, Finland, Denmark, and Canada.[4] Our faith in mobility persists in the face of such facts; without it the public rationale for a culture of individual initiative would collapse.

Tocqueville wrote:

> As social conditions become more equal, the number of persons increases who . . . owe nothing to any man, [and] expect nothing from any man; they acquire the habit of always considering themselves as standing alone, and they are apt to imagine that their whole destiny is in their own hands. Thus not only does democracy make every man forget his ancestors, but it hides his descendants and separates his contemporaries from him; it throws him back forever upon himself alone and threatens in the end to confine him entirely within the solitude of his own heart.[5]

Given our resolutely individual experience of ourselves, we should not be surprised that organized labor has collapsed, or that the social safety net is under constant pressure toward privatization, or that the company pension has given way to the Individual Retirement Account. The point is this: our cherished economic individualism has become a somewhat dysfunctional ideal, in light of the systemic inequalities that have gotten locked into our economy. Yet that ideal persists. If anything it has become more extreme as a point of reference on our political compass, and less tolerant of critique. It must certainly contribute to our experience of *individual* inadequacy, rather than collective discontent of the sort that gets expressed politically.[6]

Ehrenberg's book allows us to connect some big dots. The liberation of the individual from various identities, obligations, and allegiances in the 1960s gave a new flavor to our economic individualism. The economics of the right became infused with the moral fervor of the youthful left in a grand synthesis of liberation that gave us the figure of the bohemian entrepreneur as the exemplary human type. One effect of this trajectory has been the clinical explosion of depression (as well as a shift in how we understand our unhappiness).

Once upon a time, our problem was guilt: the feeling that you have made a mistake, with reference to something forbidden. This was felt as a stain on one's character. Ehrenberg suggests the dichotomy of the forbidden and the allowed has been replaced with an axis of the possible and the impossible. The question that hovers over your character is no longer that of how good you are, but of how capable you are, where capacity is measured in something like kilowatt hours—the raw capacity to make things happen. With this shift comes a new pathology. The affliction of guilt has given way to weariness—weariness with the vague and unending project of having to become one's fullest self. We call this depression.

Depression is especially threatening in a culture of performance, as it is a pathology in which one has difficulty initiating action. Somehow, at just the right moment in the evolution of our economic culture, we discovered in the 1980s that "a particular molecule can facilitate the ideals of autonomy, self-realization, and the ability to act by oneself," as Peter Kramer wrote in *Listening to Prozac*. We were now able to adjust ourselves to the collective demands peculiar to an assemblage of sovereign individuals.

One of the ironies of this situation is the unexpected harmony we find between a deterministic biochemical picture of the human being and the ideal of autonomy. Recall that Kant offered that ideal by way of defending the freedom of the will against material causation, by hiving it off from anything empirical and locating our freedom in a separate realm of the ideal (pp. 73–78). But the flip side of determinism is self-manipulability, and Kant doesn't seem to have anticipated the appeal that this holds for a person raised in a cultural idiom of autonomy. To regard oneself as a collection of synapses and neurotransmitters is to take a certain stance toward oneself. I don't think "I am in despair because I lost my job," I think "My serotonin levels are low, and there's a pill for that." This is to shift from a first-person perspective in which I inhabit my own experience and interpret it, giving reasons for it that refer to events in the world, to a third-person perspective in which I objectify myself and the reasons I invoke are material causes located inside my head.[7] This naturalistic determinism would have horrified Kant, but note that such inwardness gets apparent warrant from his insistence that we conceive our will as free of all those sources of heteronomy that arise from our external circumstances.

THE USES OF CONFLICT

On Freud's understanding, there is a fundamental conflict between the self and the world; that is essentially what the experience of guilt tells us. Such conflict is a source of anxiety, but it also serves to structure the individual. The project of becoming a grown-up demands that one bring one's conflicts to awareness; to intellectualize them and become articulate about them, rather than let them drive one's behavior stupidly. Being an adult involves learning to accept limits imposed by a world that doesn't fully answer to our needs; to fail at this is to remain infantile, growing old in the *Mickey Mouse Clubhouse.*

Of course, there is a hazard in the Freudian cure. There used to be a certain type: someone in interminable analysis who intellectually fetishized his conflicts and became hyperarticulate about them. Think of the characters played by Woody Allen in his early movies such as *Sleeper* and *Annie Hall.* But sometime in the late 1980s the neurotic was replaced, as a cultural type, by the depressive, who understands his unhappiness not in terms of conflict but rather in terms of mood. Mood is taken to be a function of neurotransmitters, about which there's not much to say. Inarticulacy is baked into any description of the human being that we express in neuro-talk.

Corresponding to this shift, Ehrenberg points out, is a new emphasis on well-being. In the old Freudian dispensation, to be psychoanalytically "cured" was not to achieve well-being; it was to be clear-eyed about oneself and about the human condition. Unlike many of his intellectual heirs, Freud offered a tragic view that resisted dreams of a final liberation. The interdictions of society aren't simply repression; they are formative of the kind of individual who inhabits that society. Nor is this to be understood simply as conformity. Rather, the individual is a creature who comes into being only through conflict, in some historical setting (as in Hegel). Civilization comes at a high personal cost, but the alternative would be something less than human.

Freud's thought can help to illuminate the psychological appeal of our ideal of autonomy. That ideal seems to have at its root the hope for *a self that is not in conflict with the world.*

To adopt a brain-centered perspective on oneself is perfectly suited to this hope, as it expresses the corollary hope that the self is manipulable by mood-brightening molecules that maximize well-being without reference to a person's situation: his biography, relationships, or wider cultural and economic setting.

Ehrenberg writes that the "pharmacohuman" would "no longer be subject to the usual condition known as limits." Similarly, the pioneers of virtual reality research were animated by a wish to explore the possibilities of experience without the limits that define us as humans.

What sort of self shall we choose to be? The way psychoactive drugs are currently used indicates that the "choices" we face tend to get highly funneled by societal pressures. Anecdotally it seems to be the case that, for example, junior faculty at high-powered research universities are taking as much Adderall as their students, and this is perfectly understandable. As Ehrenberg argues, a culture of self-responsibility is a culture of performance, which is a culture of competition. In light of that competition, there is really only one kind of self that is going to be successful: the high-performance kind. This starts to feel less like something chosen in a shining moment of existential freedom and more like something obligatory.

Perhaps we have merely shifted the source of our lack of freedom from identifiable external authorities (the kind one can challenge) to a net of scientistic explanations and economic pressures. Both the explanations and the pressures are predicated on an atomized picture of the self. The binding character of this net is hard to see and hard to take issue with, because it fits so comfortably. If it could speak, it would do so in the deep grammar of autonomy.

UP FROM FREEDOM

If we can put aside for a moment our centuries-long preoccupation with liberation, we might think differently about authority. The key would be to conceive authority in a way that is free of those metaphysical conceits that provoke an allergic reaction in the modern mind. Recall once more Iris Murdoch's description of learning Russian. The "authoritative structure" she invokes as a counterweight to the self is not the law of a punishing Jewish god, nor the promiscuous love of a Christian one. Rather, it is the authority of a skilled practice that "commands my respect" for reasons internal to the practice, requiring no further foundation or metaphysical support. These reasons are progressively revealed as one goes deeper into the practice.

The moral psychology Murdoch offers is entirely this-worldly. Its basic stance is one of gratitude; she speaks of "love of Russian." It is guided by a kind of pleasure: "Attention is rewarded by a knowledge of reality," she says. The role played by love in this account indicates that attention may be at bottom an erotic phenomenon.

10

THE EROTICS OF ATTENTION

In his final years, the novelist David Foster Wallace appears to have been exploring the possibility of something like mystical ecstasy, through prodigious feats of attentional self-mortification. In a note discovered after his death, he wrote:

> Bliss . . . lies on the other side of crushing, crushing boredom. Pay close attention to the most tedious thing you can find (Tax Returns, Televised Golf) and, in waves, a boredom like you've never known will wash over you and just about kill you. Ride these out, and it's like stepping from black and white into color. Like water after days in the desert. Instant bliss in every atom.

This gem reads like a report from the cutting edge of ascetic self-experimentation. The meticulous filing of tax returns and watching of televised golf are offered in the spirit of Timothy Leary's acid trips, adapted to the clean-living imperatives of someone who has struggled with chemical addiction (as Wallace did), and who perhaps retains a jones for bliss of the sort that ordinary life doesn't offer. Given the sheer effort required to stay sober, it is understandable that Wallace would be fascinated by the will, and by the transcendent possibilities of self-overcoming. "*Riding it out*, an agonized and patient abiding, seems to have been a condition of survival for much of Wallace's life," says Matt Feeney. This is the chastised, wise asceticism of a twelve-stepper rather than, say, the hectoring rectitude of Straight Edge.

Recall the quotation from Simone Weil: "Something in our soul has a far more violent repugnance for true attention than the flesh has for bodily fatigue. This something is much more closely connected with evil than is the flesh. That is why every time that we really concentrate our attention, we destroy the evil in ourselves."

Weil and Wallace both offer an ascetics of attention, whether in the service of otherworldly bliss or "destroying the evil in ourselves." Both are quasi-religious, but in a peculiarly modern way—they rely on an effort of the will rather than divine grace. I would like to offer a gentler understanding of the role that attention can play in a life, one that is entirely this-worldly. I call it an erotics of attention because the point is to fasten on objects that have intrinsic appeal, and therefore provide a source of positive energy.

FINDING VERSUS CONSTRUCTING; ATTENTION VERSUS IMAGINATION

In his commencement address at Kenyon College, Wallace suggested that

> learning how to think really means learning to exercise some control over how and what you think. It means being conscious and aware enough to choose what you pay attention to and to choose how you construct meaning from experience. Because if you cannot exercise this kind of choice in adult life, you will be totally hosed.

Wallace is saying something important: the ability to direct our attention as we will is a basic condition for living well. This sounds about two-thirds right to me, but I want to quarrel with his language of "choice"—the language of mere decision—which makes it sound like "construct[ing] meaning from experience" is somehow arbitrary, and insist rather that meaning and agency are tied in interesting ways to our efforts to reconcile ourselves to a world that is what it is, and find ways to love it.

Wallace states the central problem of life as one of critical self-awareness, as opposed to self-absorption. "[A] huge percentage of the stuff that I tend to be automatically certain of is, it turns out, totally wrong and deluded." In particular, "everything in my immediate experience supports my deep belief that I am the absolute center of the universe . . ." The task, then, is one of "somehow altering or getting free of my natural, hard-wired default setting which is to be deeply and literally self-centered and to see and interpret everything through this lens of the self." His point, he makes clear, is not a moral one about being altruistic. The point is not to be deluded, "lost in abstract argument inside my head, instead of simply paying attention to what is going on right in front of me . . ."

The reason to worry about being self-centered is that it makes it hard to cope with life. "There happen to be whole, large parts of adult American life that nobody talks about in commencement speeches. One such part involves boredom, routine and petty frustration." He describes the experience of getting off work, exhausted, and having to endure traffic and a crowded supermarket before getting home.

> The point is that petty, frustrating crap like this is exactly where the work of choosing is gonna come in. Because the traffic jams and crowded aisles and long checkout lines give me time to think, and if I don't make a conscious decision about how to think and what to pay attention to, I'm gonna be pissed and miserable every time I have to shop. Because my natural default setting is the certainty that situations like this are really all about me. About MY hungriness and MY fatigue and MY desire to just get home, and it's going to seem for all the world like everybody else is just in my way.

Wallace offers for the graduates a train of thought about fat, ugly SUV drivers, and points out that patriotic or religious bumper stickers always seem to be on the most disgustingly selfish vehicles. The crowd begins to cheer, and Wallace intervenes: "This is an example of how NOT to think, though." Then he gives some examples of what he has in mind by way of choosing to think differently.

> In this traffic, all these vehicles stopped and idling in my way, it's not impossible that some of these people in SUVs have been in horrible auto accidents in the past, and now find driving so terrifying that their therapist has all but ordered them to get a huge, heavy SUV so they can feel safe enough to drive. Or that the Hummer that just cut me off is maybe being driven by a father whose little child is hurt or sick in the seat next to him, and he's trying to get this kid to the hospital, and he's in a bigger, more legitimate hurry than I am: it is actually I who am in HIS way.[1]

Wallace concedes that "none of this is likely, but it's also not impossible." Such generosity is meant as a corrective to our default setting, which is to be sure we know what reality is, to be sure it revolves around us, and therefore not to consider "possibilities that aren't annoying and miserable" when it comes to others who stand in our way. But "if you really learn how to pay attention, then you will know there are other options."

This impatient, hostile self-absorption is spot-on as a description of my own default state, while driving especially. And Wallace is surely right about the need for charity of interpretation in our dealings with others, not least for the sake of our own tranquillity. The criticism I would like to make of his account begins with a seemingly minor point: when he suggests that the generous response results from "learn[ing] how to pay attention," I think he has misdescribed his own examples. They are acts of imagination, not attention. He is *positing* scenarios that will engage his sympathies. On this point turn some crucial matters.

The first is a practical question about how effective or sustainable such an approach is likely to be. Wallace recommends a basically Stoic strategy of minimizing one's pain by changing one's *beliefs* about the irritants that are disturbing one. The problem with the Stoic strategy is that beliefs involve states of affairs in the world, so it isn't simply up to us to decide to believe what we want. It would be nonsensical to come into a building and announce, "It's raining outside, but I don't believe it."[2] Short of such outright contradiction, one has only so much interpretive latitude before one's imaginings take on a hallucinatory aspect. Forrest Gump has a positive affect that is impervious to the world, but there is something defective about him.

Despite his repeated references to attention, Wallace's core suggestion in the speech is that "you get to consciously decide what has meaning and what doesn't." But is this not, precisely, "to see and interpret everything through this lens of the self" and thus reproduce the problem that he is trying to solve? Wallace speaks a subjectivist language in which we *posit* the world, and do so according to the free movement of our will. His solution is thus emblematic of the problem we are addressing in this book: we have an uncertain grasp of the world as something with a reality of its own. Wallace's therapy is offered in the spirit of virtual reality.

Iris Murdoch, like Wallace, is impressed by the problem of self-enclosure. But she suggests a different way out of one's head—what we might call the Epicurean way. The Epicurean recommendation, in contrast to the Stoic, is that if you are being disturbed by some unwanted emotion, it is a shift of attention, rather than a willful effort of belief, that will deliver you from it. As she writes:

> Where strong emotions of sexual love, or of hatred, resentment, or jealousy are concerned, "pure will" can usually achieve little. It is small use telling oneself "Stop being in love, stop feeling resentment, be just." What is needed is a reorientation which will provide an energy of a different kind, from a different source. Notice the metaphors of orientation and of looking . . . Deliberately falling out of love is not a jump of the will, it is the acquiring of new objects of attention and thus of new energies as a result of refocusing.[3]

Murdoch's therapy is predicated on realism: new energies come from real objects that one becomes interested in. This strikes me as more thoroughly liberating than the effort of reinterpretation that Wallace recommends. It is less concerned with moral improvement or being just. You simply *abandon* the object that is tormenting you. You walk away, and don't even notice that you have done so, because your energies are focused elsewhere. Eros is the faculty that does this for us.

Murdoch points out that "the religious person, especially if his God is conceived of as a person, is in the fortunate position of being able to focus his thought upon something which is a source of energy," and that "prayer is properly not petition, but simply an attention to God which is a form of love." She asks, "What is this attention like, and can those who are not religious believers still conceive of profiting by such an activity?"[4] This is a crucial question.

ACTING VERSUS RUMINATING

Consider as an example someone who suffers not from some raging emotion of lust, resentment, or jealousy, as in Murdoch's examples, but rather sadness, discontent, boredom, or annoyance. A wife, let us say, feels this way about her husband. But she observes a certain ritual: she says "I love you" upon retiring every night. She says this not as a report about her feelings—it is not *sincere*—but neither is it a lie. What it is is a kind of prayer. She invokes something that she values—the marital bond—and in doing so turns away from her present discontent and toward this bond, however elusive it may be as an actual experience. It has been said that ritual (as opposed to sincerity) has a "subjunctive" quality to it: one acts *as if* some state of affairs were true, or could be.[5] This would seem to be a particularly Jewish sort of wisdom—an emphasis on observance as opposed to the Protestant emphasis on inner state. It relieves one of the burden of "authenticity."

William James offers just such relief in his essay "The Gospel of Relaxation." He writes, "In order to feel kindly towards a person to whom we have been inimical, the only way is more or less deliberately to smile, to make sympathetic inquiries, and to force ourselves to say genial things . . . To wrestle with a bad feeling only pins our attention on it, and keeps it still fastened in the mind; whereas if we act *as if* from some better feeling, the old bad feeling soon folds its tent like an Arab and silently steals away."[6] We should "pay primary attention to what we do and express, and not . . . care too much for what we feel."

It might well be asked, how is the wife's subjunctive mood of prayer any different from the generous imaginings that Wallace recommends? I think the answer turns on the fact that it issues in an *action*—here, the ritual of saying "I love you" (to which it is impossible not to respond). Saying this alters somewhat the marital scene; it may not express love so much as invoke it, by incantation. One spouse invites the other to join with her in honoring the marriage, and it is the activity of doing so, together, that makes the marriage something one *could* honor. It is an act of faith: in one another, but also in a third thing, which is the marriage itself.

Likewise, if Wallace's generous imaginings of an annoying person in the supermarket checkout line were to issue in some action or utterance by Wallace that could be taken up by the stranger, becoming material for a generous response by the stranger in turn, then together Wallace and the stranger might become coauthors of a scene that is quite different from what it seemed initially, the lonely hell that Wallace describes.[7] But Wallace's generous imaginings cannot catalyze such a transformation if they remain mute and issue in no speech or deed. They are then a means of escaping the world rather than joining it.

To repeat, the Latin root of our English word "attention" is *tenere*, which means to stretch or make tense. External objects provide an attachment point for the mind; they can pull us out of ourselves. But only if they are treated *as* external objects, with a reality of their own.

SELF-PROTECTION

When someone has difficulty relating to objects (including other people) as independent things, the name for this condition is narcissism. It is not a condition of grandiosity so much as fragility;

the narcissistic personality needs constant support from the world, and is unclear on the boundary between self and other. As Sherry Turkle writes, such a personality "cannot tolerate the complex demands of other people but tries to relate to them by distorting who they are and splitting off what it needs, what it can use. So, the narcissistic self gets on with others by dealing only with their made-to-measure representations."[8]

Such representations may take the form of David Foster Wallace's generous imaginings in the supermarket checkout line, which are made to measure by Wallace for the purpose of moderating his own impatience. If these representations don't result in an interaction, they go uncontested, and Wallace is then free to "construct meaning" in whatever way best serves his psychological need. (And, Christ, maybe we should be grateful for any strategy that can prompt some humane feeling toward others.)

Another way we deal with others through representations is in the Kabuki dance of our electronic lives. Turkle conducted interviews with people about their use of various digital technologies. In her very interesting interpretation of her findings, she locates the narcissism of the e-personality not in the grandiosity of our *self*-representations, but in the simple fact that we increasingly deal with others through representations of *them* that we have. This results in interactions that are more contained, less open-ended, than a face-to-face encounter or a telephone call, giving us more control. In this domain we have a frictionless array of weak ties to other people who can be summoned according to our own needs.

You are sitting at an airport bar by yourself, feeling a bit antsy, and go through the contact list on your smartphone. You find one or two people who might appreciate the witty observation you just made, and fire off a couple of texts. Even before getting any response, you feel validated (as Turkle points out). I do this often. It is more appealing than getting bogged down in a phone call, which could go in any number of directions, and be awkward to extricate myself from if it gets stilted or boring. At such moments I am a bit like the quasi-autistic gambler who seeks control, and prefers not to deal with the full, messy presence of friends.

Armed with your list of text buddies, each of whom appreciates a particular side of your multifaceted brilliance, you also won't be called upon to respond to the person on the stool next to you at the bar. This is nice, because in such a conversation you may get an inkling—conveyed by the voice or the eyebrows—of some emotional register that was not on your agenda. Maybe he's hitting on you. Maybe he's sizing you up for some investment pitch, or getting ready to share the good news about Jesus Christ. Thank God for your phone. Then again, maybe he's just another weary traveler looking to connect, offer a wry take on the TSA, and share a chuckle.

It's not simply that we are too busy for others; we have also developed a heightened instinct for self-protection. Turkle reports that teenagers would far rather text than make a phone call because on the phone they fear that they "reveal too much." In texting you can carefully craft the version of yourself that you present.

Interviewing people about their use of social media, Turkle says her informants express "a certain fatigue with the difficulties of life with people." Real people make too many demands and constantly disappoint. She suggests we have developed a widespread emotional readiness for substitutes, for example robots that can mimic intimacy of one sort or another, as pets for the elderly, or as sexual partners for the lonely. "When people talk about relationships with robots, they talk about cheating husbands, wives who fake orgasms, and children who take drugs. They talk about how hard it is to understand family and friends."[9] There is no doubt about it: other people are a major pain in the ass. Put differently, they stand in the way of our freedom to "consciously decide what has meaning and what doesn't," to use Wallace's formula.

Faced with "how hard it is to understand family and friends," the autistic retreats into autostimulation. For his part, the narcissist splits off from others what he can use: the parts that bolster his own self-image. We recognize both as pathologies; they might also be understood as the destination toward which the ideal of autonomy tends, absent other ideals that can serve as a counterweight to it. As we saw in our discussion of Freud, the ideal of autonomy seems to have at its root the hope for *a self that is not in conflict with the world.*

One way this shows up is as an aversion to face-to-face confrontation. For all our online nastiness, my impression is that this aversion is stronger now than it was a few decades ago. This becomes apparent if you look at children's television. I have a set of DVDs of the first-generation *Sesame Street* episodes, from the late 1960s and early 1970s. Before an episode begins, a warning comes up on the screen: "This show is historical, intended for adults, and may not be suitable for viewing by today's children." And indeed the show is bracing, if you are accustomed to today's offerings. In the early 1970s, it was apparently still all right to show characters getting mad at each other. There is real conflict, as for example when Bert and Ernie are at a movie, and Ernie keeps talking during the movie, reporting to Bert his own responses to it, despite Bert's embarrassed efforts to get him to quiet down. The Muppets nearby get increasingly annoyed with Ernie. It starts with disapproving clucks and hisses, and eventually devolves into shouted insults and threats, with Muppets getting up out of their seats and coming down the aisle to get in each other's faces—a real melee.

In another episode, it is late at night and Ernie is unable to sleep. He starts singing at the window of his tenement building. The complaints start low and build into a chorus; soon there are some choice insults echoing amid the laundry lines and alley cats. You will not see anything remotely like that in today's children's programs. At some point the messy urban sociality depicted in the original *Sesame Street* gave way to a suburban scene of isolation and absolute niceness. The physical spaces depicted—the interiors of single-family dwellings—mirror the moral isolation of the autonomous liberal subject.

In one of the early episodes, a blue monster is doing what is clearly an ad-lib improvisation with two (real) children. They are eating apples together. The blue monster is conversing with them in gruff, unsoftened male tones, without any particular solicitude on display, and this seems to give their shared apple-eating a special kind of intimacy. There's not much talking, actually. Eventually the blue monster asks the children what other kinds of fruit they like. Grapes. "Uh-huh." Bananas. "Yeah." Celery. "Celery?! That's not a fruit!" He says this with unhesitating force; it is an immediate verbal slap across the face. The young boy is momentarily taken aback. But then something in his face becomes more clear. He is smiling. The blue monster takes the boy seriously enough to treat his response as a statement about the world, which can be wrong, not simply as a report about his feelings, which must be protected. In this bold bit of improvisation we witness a moment of maturation. It is a treat to watch, but it is "not suitable for viewing by children today." The tamping down of face-to-face conflict must be connected to the fragility of the contemporary self. (Meanwhile political discourse has become a performance art of fake outrage.)

On this front, consider the hipster. Christy Wampole offers us the spectacle of the tattooed twenty-five-year-old male wearing a Justin Bieber T-shirt. Or perhaps he invokes some obscure system of allusions by embracing an outmoded style (Wampole gives the example of tiny running shorts). He may take up the accordion, expressing nostalgia for an era he never lived through himself. Wampole points out that all this irony can be understood as a preemptive defense against the kind of exposure one risks in putting forward one's own aesthetic statement for others to respond to. One might be ridiculed.

Would it be possible for a rock front man like Robert Plant to appear now, after the movie *This Is Spinal Tap* has percolated through our consciousness for a couple of decades? A brilliant satire of rock, I suspect it had the unfortunate effect of helping to spawn the hipster's evasive ethic of self-protective cleverness. There is some great popular music these days, but at present it would be hard to name a band that aspires to the epochal stature of a Led Zeppelin. We seem to feel ourselves latecomers to history, as though the human story has played itself out and there remain no great deeds to be done. What is left is to play with the forms we have inherited, sampling and referencing.

In a previous chapter we considered Hegel's idea that we need other people to achieve individuality. For others to play this role for me, they have to be available to me in an unmediated way, not via a representation that is tailored to my psychic comfort. And conversely, I would have to make myself available to them in a way that puts myself at risk, not shying from a confrontation between different evaluative outlooks. For it is through such confrontations that we are pulled out of our own heads and forced to justify ourselves. In doing so, we may revise our take on things. The deepening of our understanding, and our affections, requires partners in triangulation: other people *as* other people, in relation to whom we may achieve an earned individuality of outlook.

Absent such differentiation, there is a certain flattening of the human landscape. In the next chapter, I'd like to consider how the built environment of our shared spaces may contribute to this flattening. When they are saturated with mass media, our attention is appropriated in such a way that the Public—an abstraction—comes to stand in for concrete others, and it becomes harder for us to show up for one another as individuals.

11

THE FLATTENING

I started lifting weights when I was thirteen, in the basement of the YMCA that is kitty-corner from Berkeley High. This was about 1979. The benches were covered with red vinyl that had sparkles in it, and most of the covers were worn through in spots to the foam padding underneath, leaving jagged edges of hardened vinyl right where it mattered. The foam, in turn, was dished out where head, buttocks, knees, or elbows made contact. Some had been shored up with duct tape. You got the impression that no one in an administrative capacity paid any attention to the weight room.

The mood in the low-ceilinged, windowless room was variable, but tended to take its bearings from a core group of black guys who put three or four forty-five-pound plates on each side of the bar when they were setting up for squats. They gave the impression of being linemen in permanent off-season—former contenders, keeping it together in case the call came down one last time.

There was usually a cassette player in one corner, sitting on the floor. Sometimes there was music playing, sometimes there wasn't. Sometimes the choice of music was an occasion for complaint or ridicule, which might have been answered with a reference to the size of someone's ass. Often I didn't know who had chosen the music. But it had to have been someone in the room.

Sometime around 2002 I secured privileges for using the gym at a university in the city where I live. Lying back on the bench, I wrap my hands around the bar at shoulder width and look up at the speaker in the ceiling. I believe the sounds coming out of it on this occasion are what some call "emo." On another day it might be something different, yet somehow the same. I begin to think that the recording of these sounds must have been orchestrated by some trade association of building managers; it is institutional noise of a sort that mimics music, and has become ubiquitous in public spaces such as this gym.

This strikes me as odd, especially in a university setting. Young people care a lot about their music. Right? Yet here everybody was suffering under a blanket of incontestable sonic lameness. These being the early days of the iPod, some (but not yet most) users of the gym escaped it by inserting their own earbuds.

One day I went up to the student attendant at the front desk, pointed to the audio rack by his knees, and asked, "Can you plug a CD player into that thing?" He looked a bit alarmed, and said that he didn't know. I suggested he could bring his own music. "Because anything you're into is going to be better than this." I hope I made it clear that I meant really *anything*—modern country, light FM, whatever. I guess I wanted to bring the source of the music back into the room, where it could be contested. Like a boom box visible on the floor. At the YMCA, the weight room felt as though it belonged to those who were in the room, and this place didn't.

There seemed to be a touch of panic building in the young man's eyes as I lingered at the desk, wanting to engage him on the matter of the music. I had heard someone once before ask for the music to be changed to another channel (there were several). Presumably that can be handled as a standard front desk script. But I wasn't complaining about the music per se; I was more interested to know what his role was with respect to the music. Was this role explicitly laid out? Were there hefty fines or corporal punishment prescribed for violating it, or what? Though I tried to come across as curious rather than judgmental, he probably sensed that I found something lacking in his soul. What the desk clerk said, finally, was something that really stuck in my mind. He said he didn't want to impose his choice on others.

Now, maybe he assumed I was a professor or something, and I was testing him. Maybe he thought that after he said what he said, I would say to myself, "What a principled young man," and leave him alone. In fact I did walk away, but what I was thinking, I have to confess, was "What a cow," or words to that effect.

This was unfair, of course, as he was merely playing an institutional role. Yet institutions form us, and one can't help but judge the result. This lad had been educated into an ethic of democratic niceness, and the upshot was an automatic deference to the musical programming packaged by some institutional service provider.

SUBJECTIVISM

We often talk as though aesthetic judgment is purely subjective, and therefore not the sort of thing that is amenable to public contestation. This conceit has a long pedigree: *De gustibus non est disputandum*, said the Romans. But that's not how we really feel about it. When you put forward an aesthetic judgment in public, you put yourself at risk. What you're saying is "This is *good*." That kind of full-throated affirmation has always been at odds with the agnosticism that is thought to be part of democratic good manners.

In a study conducted in the summer of 2008, the Notre Dame sociologist Christian Smith and his colleagues conducted in-depth interviews with 230 young American adults about their moral lives. What they found is nothing so exciting as depravity, but rather a depressing inarticulacy. Summarizing Smith's findings, David Brooks wrote, "Many were quick to talk about their moral feelings but hesitant to link these feelings to any broader thinking about a shared moral framework . . . As one put it, 'I mean, I guess what makes something right is how I feel about it. But different people feel different ways, so I couldn't speak on behalf of anyone else as to what's right and wrong.' "[1]

It was Thomas Hobbes who first made the privatization of judgment a political principle. Writing during the bloody English civil war, he argued not just that strong evaluations should be kept to oneself, in order to keep the peace. Not just that the public square should be denuded of "values," lest someone feel himself the object of another's disapproval. What Hobbes offered is something more radical—roughly, "What makes something right is how I feel about it," the very formula now mouthed serenely by sophomores. Call it subjectivism.

For the subjectivist, value judgments don't apprehend anything. There is no feature of the world that would make them true or false, since they merely express private feeling. It follows that your moral and aesthetic outlook can't become more discerning. It can't deepen or mature, it can only change.[2]

But surely your understanding of friendship, for example, at age forty isn't the same as it was at age thirteen. Nor is it merely different, if all goes well, but deeper. And this deepening, tied to a particular biography, is what we mean when we talk about individuality; we are referring to something that has been earned. Yet for the subjectivist, everyone is already an individual by default; everyone has his or her own idiosyncratic bundle of value sentiments. Subjectivism can't make sense of the experience of achieving greater clarity in one's evaluative outlook, and it can't accommodate the closely related idea of an earned individuality of judgment, as opposed to mere idiosyncrasy.[3]

In another context, the statement of the respondent in Smith's survey, that "what makes something right [or excellent or lame] is how I feel about it," might sound like the boast of an existential hero. But it has become the refrain of all of us who feel weak. Subjectivism leaves people isolated. Moral and aesthetic judgments have the same status as mere sensations, such as an itch—they are entirely one's own.[4] As such, they are basically incommunicable. The dogmatic inarticulacy of subjectivism—perhaps we should call it moral autism—leaves people bereft of any public language in which to express their intuitions about the better and worse, the noble and shameful, the beautiful and ugly, and *assert them as valid*.

Arguably, what it takes to be an individual is to develop a considered evaluative take on the world, and stand behind it. Doing so exposes one to conflict, and in the conversations with others that follow you may revise your take on things. Such development can't occur if you're not attached to anything to begin with, or never put it forward to others as being choiceworthy.

Further, such squeamishness creates a certain normative vacuum in our public spaces. In walking off the field of our shared moral and aesthetic life, we cede that field to corporate forces, which are not at all shy about offering up a shared experience: the emo coming out of the sound system. That's what we end up with. The way anonymous others *leap in* on our behalf and install these systems, without anyone taking responsibility for them, makes the shared experience unavailable for discussion. It can't be subject to disputation, and this is why it feels suffocating.

The taken-for-granted presence of the Muzak system spares us the exposure that comes from bringing forward one's own taste for others to respond to, as happened at the YMCA. And this process is self-reinforcing: the saturation of public space by the inevitably lame manufactured experience spurs us to plug in our earbuds, reinforcing our self-enclosure.

THE EVERYONE

As a first approximation, the gym guy was some kind of Kantian, I guess. Better to leave the Muzak undisturbed than to risk imposing his preference, which is necessarily private and arbitrary. The Muzak is at least public and neutral, seems to be the thought. Somehow it *represents* others, taken in the aggregate. And it is precisely in this aspect of averageness and anonymity that the others command deference. Such deference to an abstraction seems to be the political corollary of the subjectivist doctrine, and this makes perfect sense: surely you wouldn't want to impose your arbitrary private judgment on everyone else.

What we have, then, is a curious combination of self-aggrandizement (what makes something good is how I feel about it) and timidity. It is a combination that seems to attenuate human connection. This cannot simply be laid at the feet of Kant, because he would abhor subjectivism. In aesthetic judgment, he says, you must resist being charmed by the work of art, and not give in to your own emotional response to it, because it is possible to be *wrong* in these matters.[5] But the gym attendant's deference to the Muzak indicates that though he is likely beginning from the opposite premise from that of Kant (if he resembles the subjectivists in Smith's survey), in his public role as keeper of the music he has arrived at the same kind of self-alienation that Kant prescribes. Kant says that like moral judgment that avoids special pleading on one's own behalf, sound aesthetic judgment is "accomplished by weighing the judgment, not so much with actual, as rather with the merely possible, judgments of others, and by putting ourselves in the position of everyone else, as the result of . . . [an] abstraction from the limitations which contingently affect our own estimate."[6]

Note that Kant would have us abstract not only from our own aesthetic response, but from that of any "actual" others. We are to refer our judgment to "merely possible" others: the Everyone.

What is Muzak, if not the music of the Everyone?

Because Kant would have us put ourselves in the position of others when making aesthetic judgments, some have tried to find in this doctrine the basis for emotional connection to other people.[7] But because they cannot be concrete others, I think Kant's rigorous teachings are unlikely to get the groove going at any *actual* party.

Sometimes it is through a *contest* of individual, articulated sensibilities that the feeling of community arises. Hannah Arendt found in aesthetic disputation a glimmer of the ancient politics that she liked so much: people coming together in the public sphere and presenting them*selves*, armed not with moral rules but only with their persuasive powers, offered as evidence of the excellence of their sensibility.[8]

I wouldn't want to recommend ancient politics in general (the American founders rejected the ancient republics as a model, and for good reason). But perhaps we could preserve *some* room for this kind of agonistic yet communal feeling in our shared physical spaces.

Or maybe, as I just suggested, agonistic and *therefore* communal.

That is, precisely, how I remember the contest over the boom box at the YMCA. It is what I miss in my current experience at the gym, where the music comes from a system installed in the ceiling, controlled from a locked cabinet, and programmed by who knows who. The sense of sharing something with others is preempted or short-circuited when the music is piped in from afar and nobody takes responsibility for it. There is no *individual* with whom I could identify, or whose taste I could criticize.

To get at this difference in the two gym experiences, we need a train of thought that comes from another German thinker, Johann Gottlieb Fichte (this idea came up in our discussion of Hegel; Hegel got it from Fichte). Fichte says that individuality is born of a certain kind of interaction, in which someone issues a "summons" to you: tell me where you're coming from, in doing what you do. Give an account of yourself. In rising to this challenge, you have to own it, whatever the deed (or musical selection) is. There is an element of confrontation to it (just as when the motorcycle mechanic presents his bill to a customer).

On this view, individuality requires other people to be achieved, because it consists of a kind of setting out apart from one another. And conversely, genuine community is possible only among people who are willing to put themselves at risk in this way and *present* themselves. In doing so they may discover some fellow feeling that goes beyond politeness. (This is not at odds with my earlier plea for a "right not to be addressed," as that right applied not to individuals but rather to faceless entities that address us through mechanized means—Muzak fits this description pretty well.)

Søren Kierkegaard wrote that "only when the sense of association in society is no longer strong enough to give life to concrete realities is the Press able to create that abstraction 'the Public,' consisting of unreal individuals who never are and never can be united in an actual situation . . ."[9] Under the influence of this notion, each of us begins to view himself as a representative of something more general. We bring this "representativeness" to our encounters with others. This flattens out relationships and makes them more abstract.

Kierkegaard's concern is for the kind of setting out apart from one another that naturally occurs between people who stand in concrete relation to one another under some hierarchy, as for example between father and son, or teacher and student. Relationships like this include moments of admiring greatness, and then rebelling against its arrogance. Often there is a similar asymmetry between lover and beloved. An admirer's silent longing may have to discharge itself finally in a moment of boldness, perhaps a licentious overture. Followed by a slapped face. Genuine connection to others shows up in the vivid colors of defiance and forgiveness, reverence and rebellion, fighting and fucking: the real stuff.

The YMCA felt like "an actual situation" in a stronger way than the university gym did. By this I mean that it had more definite contours, harder edges. It would have been truly audacious for me, a skinny thirteen-year-old, to challenge the three-hundred-pound linemen for control of the boom box. But it was *thinkable* ("KBLX, really?"). The possibility hovered in the air—as a move *not* made by me. I walked right past that boom box all the time. My deference was apparent *as* deference; it was part of the atmosphere that we mutually generated in the room, each according to his place in the social dialectic, which was obviously hierarchical.

In the university gym, on the other hand, the sound system laid a blanket of abstract publicness over the situation. This flattening subtly worked against the coalescing of any genuine sociality, based on differentiation. The music was there to *represent* something: the presumed desirability of music in general.

But the music was also there to represent *me*, in a political sense: I no longer had to endure the taste of the stronger. Nor, indeed, did I have to endure the taste of some mere officeholder; the student attendant keenly felt his duty not to abuse his position. The musical situation was thus thoroughly liberal democratic. And indeed, the music itself was the music of *neutrality*. This is not really what one wants from music.

BECOMING A THIRD PARTY TO ONESELF

In the 1840s, Kierkegaard wrote that "leveling" is the victory of the public over the individual through abstraction. To conceive oneself as part of the public is flattering, in the same way that adding a bunch of zeroes after a one makes for an impressive number. Yet it is also humbling, because "even a pre-eminently gifted man . . . becomes conscious of himself as a fractional part in some quite trivial matter . . ."[10] Kierkegaard's argument is a bit obscure, but also extremely fertile, so it is worth quoting him at length on how the leveling process hollows out our relationships, leaving in their place a "colorless cohesion." He writes:

> For example, the admirer no longer cheerfully and happily acknowledges greatness, promptly expressing his appreciation, and then rebelling against its pride and arrogance. Nor is the relationship in any sense the opposite. The admirer and the object of admiration stand like two polite equals, and observe each other. A subject no longer freely honours his king or is angered at his ambition. To be a subject has come to mean something quite different; it means to be a *third party* . . . In the end the whole age becomes a committee. A father no longer curses his son in anger, using all his parental authority, nor does a son defy his father, a conflict which might end in the inwardness of forgiveness; on the contrary, their relationship is irreproachable, for it is really in process of ceasing to exist . . .[11]

Kierkegaard treats differentiating relations of authority as the incubator of genuine attachments. And these in turn make possible the individuating moment of rebellion. Kierkegaard's example of fathers and sons makes us think of contemporary quandaries in parenting. Wilfred McClay describes the "fond hope of many parents since the baby-boom generation that the painful experience of generational rebellion can be avoided in their own case—and equally, that the *agon* of individuation in their children's development can be bypassed, if only one does not make the mistake of being too authoritarian."[12] For a generation that made the rejection of authority the pillar of its permanently youthful self-image, it is intolerable to think of oneself as the object of those same rebellious passions. They hit upon the remedy of conceiving the family as a place where "parents and children are friends rather than antagonists," as McClay says. This is what Kierkegaard calls leveling. I take him to be expressing concern for the possibility of individuality—achieved through rebellion—under a kind of soft despotism that is fakely egalitarian.[13]

Leveling is intimately linked to our readiness to "go meta," to become reflective in that third-person way, and view ourselves through the lens of representation. Kierkegaard describes this:

> [We] think over the relationships of life in a higher relationship till in the end the whole generation has become a representation, who represent . . . it is difficult to say *whom*; and who think about these relationships . . . for *whose* sake it is not easy to discover. A disobedient youth is no longer in fear of his schoolmaster—the relation is rather one of indifference in which schoolmaster and pupil discuss how a good school should be run. To go to school no longer means to be in fear of the master, or merely to learn, but rather implies being *interested in the problem of education.*[14]

In this condition, our private speech comes to resemble that of a public functionary. "Nowadays one can talk with any one, and it must be admitted that people's opinions are exceedingly sensible, yet the conversation leaves one with the impression of having talked to an anonymity . . ." This was the impression left on me by the gym attendant.

When we go meta in this way, our judgments are "so objective, so all-inclusive, that it is a matter of complete indifference who expresses them . . ." Kierkegaard shows the absurd end point toward which this cultural instinct for leveling tends to move: "In Germany they even have phrasebooks for the use of lovers, and it will end with lovers sitting together talking anonymously."

To be a lover is to admire something above yourself. This becomes oppressive if it is prolonged and not reciprocated. Eros is a tyrant, a fly in the ointment of democratic morality. Eventually one becomes resentful toward the beloved, in the manner of a slave toward its master. As I suggested earlier, the moment of rebellion is that of the indecent overture, when the lover overcomes the superiority of the beloved's beauty, or vaunts himself over the weakness inflicted on him (or her) by his (or her) own desire. In doing this he incurs a genuine risk; he will probably be shot down, mid-vaunt. That is, it will be revealed to him that the vision of intimacy he entertained was an unrealistic fantasy because—guess what? He is not qualified to play that role with this person. What was he *thinking*? The risk taken in the sexual overture is perhaps a higher-stakes version of the risk taken by the DJ at a party or club. He offers himself up with his playlist in the hope that others will find something lovable, resulting in some common feeling and a genuine connection.

I suspect the depressing picture Kierkegaard presents, of men and women sitting together exchanging approved formulas out of a phrase book for lovers, was intended as a joke at the expense of the Germans. But jokes by philosophers have a way of becoming less funny as the absurdities they have identified blossom more fully, and become social norms. The joke becomes a simple description. Kierkegaard seems to have anticipated the kind of relationship talk we are encouraged to adopt in an administrative atmosphere of sexual "transparency."

The really Germanic moment of American sexuality, at least on campus, probably occurred sometime in the early 1990s. If you recall, at that time the Office of the Dean for Student Life, and various entities with names like that, encouraged students to exchange mutual statements of explicit consent before touching each other in certain clearly demarcated zones, like a TSA agent getting ready to pat you down. In the administrators' vision, men and women were to draw near and address one another as "Citizen"—each a representative of the ideal, autonomous liberal subject—and commence a negotiation that would pass muster, should the books be examined later.

In the last decade and a half, the Great Pornification has presented a different sort of script for men and women, a different way to be representatives. The forms and exchanges of porn that now stock the imaginative repertoire of those who grew up with it—including, for example, the "standard porn noises" that a friend of mine complained his lover made—perhaps alleviate the quandary of real intimacy. Like the script offered by the campus functionaries, porn solves a problem and mitigates a risk: the trackless leap that one takes in offering up one's full presence—one's defining desire, naked of social forms—to another person, not quite knowing what that person's response will be. Such exposure is felt as a moment of acute heteronomy. It's hard to be self-possessed when what you desperately want is for your most illegal desire to be desired, your most wicked self embraced, precisely for its wickedness.

The merest acquaintance with online porn—with the inexhaustible drop-down menus of sexual options—destroys any sense you might have had that in some outré sexual adventure your soul is at stake, and damnation nearby. Whatever you're into, there's a website for that. We all play "a fractional part in some quite trivial matter," as Kierkegaard says. And this is soothing. In the current dispensation, self-possession is safe from being discomfited by erotic longing.

The gym attendant was enacting a venerable norm of democratic institutions by effacing himself, the better to represent others. Kierkegaard's arresting thought is that the effacing of the individual under the banner of representation tends to work its way inward from public life to our private relationships.

This must hold some psychological appeal for us; it must answer to some need. Suppose Kierkegaard is right that genuine connection to others depends on differentiation: one encounters another person as a concrete other, for example in sexual intimacy. Or in the way a student encounters an important teacher—here too, one experiences oneself as incomplete, and therefore in need of the other. One presents oneself boldly (that is, exposes oneself as needy), submits to the other's cold gaze, and hopes that it will warm.

This threatens a self whose dignity is based on the idea of self-responsibility or self-sufficiency. And therein lies the appeal of viewing oneself, as well as others, as representatives of something general. There is then no complementarity between us, no differentiation and dependence, but instead a "colorless cohesion" of interchangeable, autonomous subjects.

This is a de-eroticized sort of gaze to direct at one another. The ecology of attention that prevails among persons in a liberal public culture is one of polite separation.

12

THE STATISTICAL SELF

It would be interesting to know if Kierkegaard had read Alexis de Tocqueville's *Democracy in America,* because in that book too we find a counterintuitive connection between the ideal of standing alone and the tendency to regard oneself as a "representative." Tocqueville was struck by the observation that as hyper-Protestants who reject anything that looks like clerical authority, Americans are expected to be self-sufficient in forming their own judgments about everything. This isn't understood as a rare accomplishment, or a capacity that one grows into in the course of a life. It is a moral imperative from the get-go, taught in elementary school.

But of course we run into a problem: we are not competent to judge everything for ourselves. We know this; we feel it. We cannot look to custom or established authority, so we look around to see what *everyone else* thinks. The demand to be an individual makes us feel anxious, and the remedy for this, ironically enough, is conformity. We become more deferential to public opinion.

Here is an example that seems to fit Tocqueville's insight. The Kinsey Reports on Americans' sexual practices became objects of intense popular interest, maybe because they arrived (in 1948 and 1953) just as the received norms and mores were loosing their grip. Everyone was left to his or her own devices. People wanted to know if they were "normal," where the only norms available, the only ones not discredited as "repression" by the pop Freudianism that swept America after the war, were now quantitative. How often do other couples have sex? What's the *average*? Is oral sex something that is done by most people? I like to be tied up—am I sick? The normative center of gravity now resides in the middle of a distribution, rather than coming from a religious interdiction or parental guidance, on the one hand, or from a cultivated, proudly antinomian sense of oneself as a pervert and sinner, on the other. This seems to fit Kierkegaard's rubric of "the death of rebellion," which is a corollary of the death of reverence and obedience. One takes one's bearings from the "representative," in the sense statisticians give to that word.

THE AVERAGED AMERICAN

The birth of social surveys is a fascinating story told by Sarah Igo in *The Averaged American*.[1] Though she doesn't mention Tocqueville or Kierkegaard, Igo offers a precise and detailed account of the paradoxical cultural logic whereby the ideal of autonomy prepares the way for massification.

The arrival of the first Kinsey Report was a very big deal in 1948. It was "a revolution in the facts Americans knew about one another," as one commenter put it. The status of these revelations as social scientific facts depended on Kinsey's assertions of the representativeness of his sample, and likewise, challenges to the report from various quarters, some of them quite alarmed, often took aim at precisely this assertion. On all sides, there was great emotional investment in the question of how representative the respondents in Kinsey's study were.

Kinsey himself was an interesting character. He had previously been an entomologist, and relied on this fact to present himself as a man of science who just happened to have turned his disinterested gaze from beetles to human sexuality. But in fact Kinsey, a respectable professor in the Midwest, had sexual tastes that weren't very conventional, and his reports appear to have been motivated by a desire to reconcile these two facts. The psychic force driving his efforts came from his hatred of "hypocrisy," and a desire to liberate the world sexually. Liberation apparently required sincerity, bringing every hidden thing out into the light.

After the first report was published, people were eager to enroll themselves as subjects for the second study; there was a long waiting list. Igo writes that "within Kinsey's vast correspondence can be found thousands of individuals seeking and finding statistical reassurance." The psychic benefit sought, and apparently found, by participating in these surveys, Igo writes, was "membership in a community of potentially similar, though anonymous, others." The reports thus served a "mass psychotherapeutic function," and Kinsey embraced his unofficial role as therapist. The format of the survey structured a respondent's identification with these anonymous others by way of categories such as black, professional, upper-class, educated (to list the tags chosen by a woman who wrote in *Ebony* about her experience of being interviewed by Kinsey's team). Igo writes that "individuals were coming to view themselves through the social scientific categories [Kinsey] and others had made available."

This new mode of self-understanding was made poignantly clear by the many instances in which participants would write to Kinsey later to follow up, update, elaborate, and correct the sexual histories and behaviors they had presented during their interview (which nominally lasted two hours, but often stretched far longer). Igo suggests many participants came to view their Kinsey interview as their "real" history—as the truth about themselves, which was otherwise elusive—and therefore they wanted the record to be complete. There was apparently a certain freedom in narrating oneself that came from knowing that, as one participant reported, "you are not making an impression, only a statistic." That is, you are not registering as an individual, and are therefore relieved of the embarrassment that might be natural to a conversation with a stranger about your sexual behavior. Another participant said that in being interviewed by Kinsey, "it seems, after a while, like you're talking about someone else instead of yourself."

Igo suggests that in having us talk about ourselves in this

way, social surveys played an educative role. They prompted Americans to take an "objective" stance toward their own lives; to view their experiences "like specimens, using the social scientist's words instead of their own to tell themselves who they were." Such a "neutral vocabulary," as the pollster George Gallup called it, encouraged those who adopted it to become third parties to themselves, to use Kierkegaard's formula.

One transvestite wrote to Kinsey to offer his sexual history with the preamble "You will find my case history somewhat typical." Presumably this was an attempt to appear sophisticated when writing to the famous social scientist. What sophistication consists in here is not being overly impressed with a sense of your own singularity.

The story of survey research begins in the 1920s and 1930s with George Gallup and Elmo Roper, names still familiar to us from the reports that continue to issue from their firms. Polling was presented by its champions as a means of "constructing rational citizens," Igo writes, and of making democracy more democratic. As Europe slid into fascism and communism, Gallup in particular was very energetic in publicizing his faith in the people to make good decisions. Igo writes that "pollsters' democratic rhetoric relied on a notion of individuals able to speak and know their own minds." But to speak and know their own minds, people first have to be removed from their social setting (just as in the Lockean thought experiment about a "state of nature" that we considered in the Interlude: a state where there is no such thing as authority and one obeys only the dictates of one's own reason).

Gallup and Roper forged "a purely statistical public from groups of randomly selected strangers." This aggregation of isolated strangers was something new, very different from any actual community. Igo writes that the pollsters "employed scientific sampling to better hear 'the man in the street' but instead created an averaged-out and abstracted public opinion that severed attitudes from their source."

The "rational" citizen is apparently a decontextualized citizen, precisely the opposite of the situated self. The pollsters' educative program matches Kant's exhortations to view oneself as a representative of the generic category "rational being." The anthropology that is tacit in polling also resonated with cultural currents first identified by Tocqueville, for example Americans' readiness to pick up and move to someplace far away, where they don't know anybody. In an influential 1991 study, social psychologists who study differences across cultures note the "abstract, situation-free self-descriptions that form the core of the American, independent self-concept," as compared with the Japanese, for example.[2]

In the somewhat self-aggrandizing rhetoric of the pollsters, their work of helping respondents to know their own minds by eliciting their views in a setting free of social pressures empowered Americans against the rise of various would-be authoritarians of the left and right. This note of democratic valor resounded in Kinsey's project too, where the authoritarians in question were various species of sexual moralists. But this liberation from the kind of cultural authority that operates in actual communities—in religions, in locales, in families—seems to lead to a feeling of isolation, which can be alleviated only by discovering that one is "normal." And so the expert of normalcy becomes the new priest, salving our souls with the offer of statistical communion.

An especially perceptive contemporary commenter on the Kinsey Reports, Lionel Trilling, suggested that they established "the community of sexuality," a community that we discover in the numbers. "We must assure ourselves by statistical science that the solitude is imaginary," Trilling wrote. The numbers offer solace.

THE STACKABLE SELF

But now the story of surveys takes an interesting twist. They didn't simply have a homogenizing effect; through abstraction and reaggregation they could also have a *differentiating* effect. In doing marketing research, Igo writes, polling firms eventually discovered that "their object was nothing so vague as 'the public,' but . . . more focused demographic groups." Likewise, social statistics "prompted some to imagine themselves into new collectives or forge a minority consciousness." We might call these the first virtual communities, composed of individuals who are spatially separated and do not know one another. For example, Kinsey's data on the prevalence of homosexuality became a tool in the movement for gay rights, and a sort of epistemic foundation for gay identity politics. Igo writes that social scientific data "created novel possibilities for community and self-assertion even as they placed new constraints on self-fashioning." Those new constraints on self-fashioning arose from the fact that, in their maturity, social surveys became the educative basis not for a mass society, but rather for a society in which people were assigned to various boxes. This encouraged "new links between strangers even as it eroded older bonds of family, religion, and locale."[3] As the inhabitant of a family, religion, or locale, a gay person was likely to stay in the closet. With the rise of identity politics, one jumped out of the closet and into the box.

The stackable self is evidently one that is especially receptive to the categories of self-understanding offered by social science, which may or may not be more confining than the communal ones they displace. To assume that they are in every case less distorting of a person's lived experience is to make an assumption that is worth revisiting in light of the arguments developed in this book: that settled forms of social authority act only as impediments to the authentic self. This assumption tends to be accompanied by another: that science is inherently liberating, and therefore in the service of this authentic self.

Igo notes that Kinsey "sought to uncover how ordinary Americans actually behaved in their sexual lives so as to liberate them from social conventions, but one of the key consequences of his *Sexual Behavior* studies, and the national discussion surrounding them, was the public shaping of 'normal' private selves."[4] Kinsey sought to remove the stigma from sexual practices thought to be deviant by showing their prevalence. The effect of this publicity was to make no place safe from the idea of normalcy.[5]

INDIVIDUALITY IS PASSÉ

Merely to raise concern about the fate of individuality in contemporary culture is probably to appear old-fashioned. It would seem to be a 1950s through 1970s sort of preoccupation, the stuff of *The Catcher in the Rye* or the cigarette and hi-fi advertisements in your dad's old *Playboys*. "Conformity" was the great worry of half a century ago.

Now we are fascinated with "the wisdom of crowds" and "the hive mind." We are told that there is a superior global intelligence arising in the Web itself. This collective mind is more meta, more synoptic and synthetic, than any one of us, and aren't these the defining features of intelligence? Of course all this crowd-loving lines up pretty well with Silicon Valley's distaste for the concept of intellectual property, and with the fact that there is a lot more money to be made as an aggregator of "content" than as a producer of it. (It is the aggregator who controls advertisers' access to consumers' eyeballs.)

"Ideology" could be taken (somewhat narrowly) to mean an idea that happens to line up with the material interests of those who espouse it. The alignment gives added psychic force to the idea, all the more so if the champions of the idea remain unselfconscious about this connection. Their enthusiasm tends to reverberate outward, and is adopted by others who have no interests at stake and therefore look naive for adopting it.

I sometimes go to the library at a well-regarded local university to write. Like most universities these days, this one is very diligent about locating itself at the cutting edge of every trend. When you get a cup of coffee, it comes with a sleeve. On this sleeve, the university takes the opportunity to profile student success stories. Recently I got a sleeve with a picture of a student in the continuing studies program. The caption read, "A master's degree allowed her to progress from writer to content expert." Apparently the young woman "progressed" from being a writer to someone who aggregates bits of other people's writing. To me that sounds more like defeat. In countless little ways, any single one of which seems trivial, this liberal arts college is unthinkingly repeating bits of Silicon Valley ideology that would seem to undermine the rationale for studying the liberal arts. The university has become "the brilliant ally of its own gravediggers," to borrow a phrase from Milan Kundera.[6]

Jaron Lanier criticizes what he calls "digital Maoism," a "new online collectivism" that shows up, for example, in the way Wikipedia is regarded and used, and is the guiding spirit of firms such as Google as well. The analogy with Maoism is quite apt and precise. The ideologists of the Web have always been antielitists, eager to brush the "gatekeepers" of knowledge into the dustbin of history. Let a thousand flowers bloom. The problem, of course, is that it's hard for these leaders of the people to make money off scattered flowers. Better to "have influence concentrated in a bottleneck that can channel the collective with the most verity and force," Lanier writes.[7] The Party must be strong for the People to be strong.

Writing about the Web in 2006, Lanier said that "in the last year or two the trend has been to remove the scent of people, so as to come as close as possible to simulating the appearance of content emerging out of the Web as if it were speaking to us as a supernatural oracle." He was referring to "consensus Web filters" that assemble material from other sites that are themselves aggregators of other sites. "We are now reading what a collectivity algorithm derives from what other collectivity algorithms derived from what collectives chose from what a population of mostly amateur writers wrote anonymously."

Lanier points out that these developments aren't confined to online culture. The elevation of the collective through the fetish of aggregation is "having a profound influence on how decisions are made in America," in government agencies, corporate planning departments, and universities. He reports that, as a consultant, he used to be asked to "test an idea or propose a new one to solve a problem. In the last couple years I've been asked to work quite differently. You might find me and the other consultants filling out survey forms or tweaking edits to a collective essay."

Lanier suggests there are institutional reasons for the appeal of collectivism in large organizations: "If the principle is correct, then individuals should not be required to take on risks or responsibilities." This is especially attractive given that "we live in times of tremendous uncertainties coupled with infinite liability phobia, and we must function within institutions that are loyal to no executive, much less to any lower level member. Every individual who is afraid to say the wrong thing within his or her organization is safer when hiding behind a wiki or some other Meta aggregation ritual."

In his own participation in such rituals, Lanier reports that "what I've seen is a loss of insight and subtlety, a disregard for the nuances of considered opinions, and an increased tendency to enshrine the official or normative beliefs of an institution."

Let's put this development in a larger context, the better to see its continuity with the cultural logic we have identified in other settings. With the Reagan/Thatcher revolution of the 1980s, the figure of the entrepreneur came to be central to our economic self-image. Individual initiative was the measure of personal value, and the hierarchical business firm came to be derided as hidebound in any number of business bestsellers. The new ideal was that every employee, from top to bottom, should have the entrepreneurial spirit, and display the virtues of autonomous behavior. Of course, employees now faced the hazards of entrepreneurship as well: heightened competition with one another, and indeed with workers in distant countries. The concept of loyalty was replaced with mobility. The expectation of continuity—of having a career, based on the steady accumulation of experience and expertise—was revealed as nothing more than cowardice. The narrative arc of work was dissolved into the isolated moments of an eternal present, each equally fraught with opportunity and insecurity.

This atomization of workers in the eighties through the aughts likely prepared the way for the new collectivism that Lanier has identified. Of course, this sounds paradoxical. But it is a paradox that lies at the very heart of individualism, identified by Tocqueville long ago as he traveled around America.

One thing that Tocqueville saw when he came to America was high levels of mobility and opportunity. This is part of what he called the "democratic social condition," which we considered earlier. For all its benefits, it brings with it insecurity: you can fall as well as rise. In the relatively rigid social systems of Europe, there wasn't much prospect of one's fortunes altering in any decisive way. This offered a certain amount of freedom from the need to manage appearances. Appearances didn't have to be generated anew each day by performing one's social value, or voicing the correct opinions. In Europe Tocqueville saw greater freedom of thought, a greater diversity of human types, and, for all the limitations imposed on the press, less prudish self-censorship among writers.

Our concept of a self that is mobile because not situated or limited in any decisive way by its current circumstances—the decontextualized "rational citizen" idealized by the polling firms, and by Locke and Kant—is tied to a certain picture of human agency: action "has no other source than the agent who accomplishes it and who takes sole responsibility for it," as Ehrenberg put it. This individualist view of action is hard to square with the experience of, say, working as a middle manager at the corporate headquarters of Best Buy. In such a setting, the chain of cause and effect is likely to be fairly opaque to you; there are issues in play and decisions being made that you are clueless about, because they are dealt with higher up in the food chain. In *Shop Class as Soulcraft* I reported the sociological finding that workers in such a position find ingenious ways to avoid taking responsibility—mainly, by making their language as vague and empty as possible, so as to preserve for themselves a maximum flexibility to reinterpret their utterances retroactively, should the circumstances demand it. Apparently, the new strategy is to hide behind "a wiki or some other Meta aggregation ritual," as Lanier says. And indeed it makes good sense to shy from offering up, and standing behind, your thoughts and utterances *as* your own if doing so could result in having something pinned on you. Instead one mouths the currently prevailing view of things while putting on the necessary dramatic performance—of individual initiative.

Note how well these developments fit with Tocqueville's idea that the massification of the American mind is a direct response to the burden of individual responsibility; to the feeling that you have been cut off from identifiable, responsible sources of authority outside yourself and must stand alone without guidance or support. Under this condition you take shelter wherever you can, and there is safety in numbers. But now you find that you have become subject to an amorphous form of authority: a gray fog that emanates from the collective, which nobody takes responsibility for.

It is hard to see where this fog is coming from. It is hard to avoid it, and hard to take issue with it. Kind of like the music coming out of speakers in the ceiling.

LOOKING PAST THE PRESENT

When the sovereignty of the self requires that the inheritance of the past be disqualified as a guide to action and meaning, we confine ourselves in an eternal present. If subjectivism works against the coalescing of communities and traditions in which genuine individuals can arise, does the opposite follow? Do communities that look to established forms for the meanings of things somehow cultivate individuality? This is the theme of the next section.

But here we come up against a methodological problem. On the one hand, to speak about "community" in general is to be led almost necessarily into idealistic blather. This would not be very informative, and would also tend to alarm some people: those who maintain the enlightener's vigilance against the threat that communal authority poses to individual self-fashioning. It would be easy to trigger this defensive reflex while also tickling a contrary sentimental reflex among those who long for "lost community." But I don't want merely to press PLAY on a dusty old culture war cassette.

On the other hand, to avoid generalities and go deep into the particulars of some community of skilled practice requires getting very technical, because it is precisely the details that the members themselves care about, and we are trying to understand what moves them. The problem is that I can assume in the reader only so much tolerance for the history and technical details of Baroque pipe organs. Organ making is not a topic that I had the least interest in myself before stumbling upon a group of people who happen to have made these instruments the focus of their working lives.

In the pages that follow I have tried to manage this dilemma, on the assumption that the reader is interested in the ways an inheritance can situate the self, and facilitate the development of an earned independence of judgment. Whether the following account avoids the pitfalls of "nostalgia" you will have to judge yourself. In any case, the sentimentalism that any depiction of craftwork naturally evokes is, I believe, only the most superficial layer of what is appealing in the scenes I have tried to capture.

PART III

INHERITANCE

13

THE ORGAN MAKERS' SHOP

Some of the best pipe organs in the world are made by George Taylor and John Boody and their team of craftspeople in Virginia's Shenandoah Valley. It is a business in which the employees require long acculturation into the history and finer points of the trade. They are able to trace lineages of who taught whom in the overlapping networks of apprenticeship among shops that do similar work around the world. In this fraternity, which includes people living and others long dead, the spirit of emulation and rivalry is intense; they try to outdo one another in making the best organs possible. The work is historically and socially situated in this way, and seems to invite each of its practitioners to experience his or her own development as a craftsperson as a chapter in a longer historical arc.

In the United States (but not Germany, for example), the idea of apprenticeship is criticized for being too narrow an education. It is said that what the economy demands is workers who are flexible. The ideal seems to be that they shouldn't be burdened with any particular set of skills or knowledge; what is wanted is a generic smartness, the kind one is certified to have by admission to an elite university. This fits well with our ideal of the unencumbered self, and with Kant's exhortation to view ourselves under the generic heading "rational being." We are told the economy is in a state of radical flux; "disruption" is spoken of as though it were a measure of value creation, and so a twenty-first-century education must form workers into material that is similarly indeterminate and disruptable. The less situated, the better.

But consider that when you go deep into some particular skill or art, it trains your powers of concentration and perception. You become more discerning about the objects you are dealing with and, if all goes well, begin to care viscerally about quality, because you have been initiated into an ethic of caring about what you are doing. Usually this happens by the example of some particular person, a mentor, who exemplifies that spirit of craftsmanship. You hear disgust in his voice, or see pleasure on his face, in response to some detail that would be literally invisible to someone not initiated. In this way, judgment develops alongside emotional involvement, unified in what Polanyi calls personal knowledge. Technical training in such a setting, though narrow in its immediate application, may be understood as part of education in the broadest sense: intellectual and moral formation.

Technologists who work in a long tradition with inherited forms also offer a useful contrast to our current image of the innovator-entrepreneur as a sort of existential hero who creates the New ex nihilo. After a period of solitary gestation in a California garage, he emerges to disrupt us and deliver us.[1]

What emerged in my conversations at Taylor and Boody is that the historical inheritance of a long tradition of organ making seems not to burden these craftspeople, but rather to energize their efforts in innovation. They intend for their organs still to be in use four hundred years from now, and this orientation toward the future requires a critical engagement with the designs and building methods of the past. They learn from the past masters, interrogate their wisdom, and push the conversation *further* in an ongoing dialectic of reverence and rebellion. Their own progress in skill and understanding is thus a contribution to something larger; their earned independence of judgment represents a deepening of the craft itself. This is a story about the progressive possibilities of tradition, then.

It is a story that is relevant to our current economic moment. With global labor markets and progress in automation, the wealthy countries will surely never again have mass employment in manufacturing; I am not suggesting that is in the cards. But there are indications that we are on the cusp of a new renaissance of small-batch, specialty manufacturing in the United States, and probably in other places too.

It would be hard to overstate the excitement you hear in people's voices when they start talking about some of the new digital tools that have drastically reduced the cost of prototyping (and some of these tools are used at Taylor and Boody, despite their antiquarian image). Design ideas can be turned into real things, and tried out, without huge financial risk. This plays to the strengths of tinkerers and inventors, those erstwhile American types who may become prominent once again. Ironically, a decades-old pipe organ shop in rural Virginia, which is caught up in a conversation with earlier centuries, may offer some guidance for the *new* "new economy."

TAYLOR AND BOODY

Pipe organs were to the Baroque era what the Apollo moon rockets were to the 1960s: enormously complex machines that focused the gaze of a people upward. Pushing the envelope of the engineering arts, a finished organ stood as a monument of knowledge and cooperation. Installed in the spiritual center of a town, a pipe organ mimics the human voice on a more powerful scale, and summons a congregation to join their voices to it. The point is to praise something glorious that transcends man's making. Yet the congregants can't help but notice that this music of praise, like the instrument that carries it aloft, is itself glorious.

A big pipe organ thus expresses both humble piety and vaunting pride at once. It can be shockingly indiscreet in this latter role; the organ often dwarfs the ostensible altar. But perhaps these tendencies get blurred together in the life of a congregation. When the choir is at full song, the stained glass is rattling loose, and the whole house seems ready to launch, what then? Then the organist pulls out all the stops. He shifts his weight to the right. His left foot is poised over the leftmost pedal, the low C, and now he stomps it, sending a thousand cubic feet of air per minute through massive pipes to blast heaven's favorite pigeons out of the rafters. Now the very pews transmit joy to women's loins, and the strongest man in the congregation feels himself reduced to a blushing bride of Christ. Now one feels it is God's own organ that fills the sacred chamber, and when this happens, praise comes naturally: hallelujah![2]

To be the maker of such an organ, a man must have a bit of sacrilege in him. Yet he must also have something like reverence, as the pipe organ comes to us through tradition. Such are the paradoxes of the organ maker.

For my appointment with John Boody, I rode west from Richmond to Afton on Route 6, then found my way to Hebron Road near Staunton, a narrow ribbon of blacktop that winds through the green cattle farms of Augusta County, to an area drained by Eidson Creek.[3] Taylor and Boody is located in a former schoolhouse that stands atop a rise in the land, directly across from an ancient cemetery and church, Hebron Presbyterian. They make organs from scratch, starting with locally felled trees for their sawmill and ingots of lead and tin for their foundry. Sixteen people work there. Their most elaborate organs cost in the ballpark of two million dollars apiece, and business is good.

The cemetery's stone boundary marker, dated 1746, had me wondering—are people buried here on Hebron Road who knew of Bach as a contemporary, and might have anticipated one of his works for the pipe organ as a new release? I pulled open the outer door of the shop building to reveal an alcove suffused with an even-tempered daylight; a short flight of wooden stairs creaked as I mounted them to an inner set of doors. In certain old buildings one feels a patina of use and settled purpose that strikes one sensually. One feels already oriented, as though a trace has been left by the movements of people engaged in some steady activity.

The inner set of doors opened into a very large undivided space. The entire west-facing wall was windows looking out onto rolling green farmland. Beneath these windows were workbenches that looked as though they had grown there; atop the benches lay planes and chisels and all the hand tools of the cabinetmaker's art. In this room there was something like a wooden space shuttle taking shape in outline—the case work for Opus 57, destined for the First Presbyterian Church in Pittsford, New York. Like composers, the makers of organs designate their works by sequential opus number. I hadn't time to take it in before a mustachioed guy holding a chisel asked me, "Are you the clavichord maker?"

I had to reply that, whatever a clavichord is, I'm not the guy who makes them. "Is John around? I'm a bit early."

"Through those doors."

Through the doors was a smaller room, around the perimeter of which were heavy floor-standing mortisers, drill presses, grinders, layout tables, and some other implements I couldn't identify. There were two men, one probably about forty and long-haired, the other in his fifties and short-haired.

"Hey. I'm looking for John."

"Boody-Man? Through those doors. The orange ones," the younger one said.

The orange doors opened into another vast room, this one a modern addition to the schoolhouse. Open to the sunlight on one side through a rolled-up bay door, it had all the equipment of a production woodshop. A dust collection system ran through galvanized duct to each machine. John Boody, who appears to be about sixty, was wearing earplugs, jeans, and a blue T-shirt, standing before a 1960s-era Delta twelve-inch table saw with a big, beautiful piece of walnut in his hand. He noticed me and waved, then proceeded to rip the walnut. He pushed it halfway across the saw, walked around to the outfeed table, pulled it the rest of the way through, and turned off the machine. Then he came over and greeted me cheerfully, still wearing his earplugs.

"I'm early, so if you're in the middle of something, I can wait," I said loudly.

"Nah."

He took a coiled air line from its perch on the wall and blew the black sawdust off his clothes. "You want to see some pipe making?" "Yes." There would be no preliminaries in this conversation. In the pipe shop John introduced me to Jeff Peterson, a reticent man of about fifty, with long hair, tattoos, and the vibe of an old-school biker. His tool cabinet sported some Harley insignia and a bodacious swimsuit model. Jeff was using a type of wood plane called a cabinet scraper, drawing it toward him in long, even strokes. But he wasn't scraping wood, he was scraping a sheet made of a very particular mix of lead and tin. The metal peeled off in delicate strips, similar to wood chips but not curly. He handed me one; it was soft. John explained that the tin and lead had arrived as solid ingots, were melted in the shop's furnace, mixed to the proper ratio, and poured into long sheets on the casting table, tapered in thickness. After this foundry work, the sheets were pounded with a drop hammer, a relic machine made in the nineteenth century that sits next to the casting table. The pounding anneals the metal, making it malleable. The scraping Jeff was currently doing gives it a certain historically correct look. Meanwhile Robbie Lawson, in his thirties and clean-cut, sat at a workbench with a printout of numbers, a scribing compass, and sheets of this proprietary metal. He gauged their thickness with a deep-throated dial micrometer, then scribed sections of annuli (doughnuts) on them: truncated cones as unfolded onto two dimensions. These lines were for cutting out sections of sheet metal, which he then bent around a cone-shaped wooden mandrel. Finally, each section was soldered together edgewise by Jeff to form the tapered toe of an organ pipe. Robbie worked as a mechanic at a Volkswagen shop before coming to work at Taylor and Boody in 1996. Jeff began as a pipe maker at the Rodgers Organ Company in Eugene, Oregon, and has been at Taylor and Boody about seventeen years.

The pipes they were making on this day were for a restoration job: an organ built in 1830 by Henry Erben. The lowest three pipes of this organ had been abducted, "borrowed" for another organ at some point in the last 170-odd years, and though they were recovered, in the course of this misadventure their windways had been altered and their ends chopped off. Some of the other pipes had gone brittle, and their ends had crumbled, requiring extensions to be soldered on. The entire Clarionet stop was missing. (A stop is a group of pipes that get activated when an organist "pulls out the stop": one of those knobs that you see on an organ console. These groups of pipes are often meant to mimic the sound of some particular instrument, in this case the Clarionet, or clarinet. Some of these instruments are extinct, and exist now only as organ stops.) John explained that this was fairly typical, and that European organs have had the worst of it, especially during World War II when their pipes sometimes got melted down for munitions and many churches were bombed. In repairing the deteriorated pipes and fabricating the missing ones, Jeff and Robbie had duplicated the details of the intact pipes, copying the Clarionet pipes from another Erben organ. John offers as a rough rule of thumb that restoring an organ costs about twice as much as building one from scratch.

THE ENTREPRENEUR

From the pipe shop we made our way to John's office, a choice space in the old part of the building. In the corner stood a small woodstove; in a rack above his desk were tubes of architectural-looking drawings. Another rack, at floor level, held white pattern boards dense with inscrutable markings and holes, and lettering in various languages. John's desk was a purposeful riot of tools and books, both antique and modern, all evidently in use. A block plane lay next to an unopened pack of Bosch jigsaw blades; a catalog of modern woodworking machines lay next to a facsimile edition of Dom Bedos's *The Organ Builder.* Illustrated with foldout drawings of exquisite mechanical detail, this work was originally published as part of Diderot's *Encylopédie*. John Boody reads technical treatises written in German and Swedish as well as English, some printed in a Gothic script requiring paleographic skills. A scholar and a musician as well as an artisan, he presents the image of a humanist from another era.

John offered me a seat. Asked how he got into organ making, he says he took every technical class his high school offered. Then he went to university as a music major, specializing in voice. His trajectory prepared him, then, to fall in love with the organ both as a musical instrument and as something to be fabricated. "In my freshman year somebody gave me a small organ, and it was all over." Here was a field that would stretch his ingenuity and give it aesthetic focus. But how did an instrument associated with ancient Saxon cathedrals come to be made in Staunton, Virginia? For Taylor and Boody build primarily Baroque organs, and hold a prominent place in what organ builders call the Baroque revival, or more broadly the organ reform movement. I asked John how the business got started.

He and his partner George Taylor had worked together for almost seven years at the organ shop of John Brombaugh in Ohio. Brombaugh, in turn, had apprenticed with the German organ maker Rudolf von Beckerath. Speaking of his time at Brombaugh's shop, Boody said, "We built about thirty organs in that partnership. John's a totally creative person, and had this concept of returning organ building to its historic base instruments. He was fascinated by these sixteenth- and seventeenth-century organs in northern Germany, so he went around and studied them. John did this whole *return*, with George and me and a couple of other partners, a five-way partnership, which is a horrible thing to do. Really one boss, but five people to share all the financial loss. One year I think we made thirty cents an hour. But we shaped up this concept of returning to the historic principles of organ building and authentic construction, and we practiced it."

Eventually George Taylor and John Boody struck out on their own. "Brombaugh gave us one contract, and that's our Opus 2. We built it in the garage behind my house. In '79 we decided we didn't want to be close to, you know, a big steel city, Middletown. It wasn't very nice. We came down here and looked around and found this building. It was like wagons going east. We loaded all the stuff and all the families and all the furniture, all the lumber and all the tools, the machinery, and we moved down here and we spent nine months renovating this building. Since then, this is our home, and we've never looked back, as they say. Well, we may have, but it was too late. And now I have my son Erik working here." He also has a number of workers who started learning their trade at other organ makers' shops, including the one where Taylor and Boody themselves started forty years ago. The making of historically inspired organs seems to be the business of a community, one that is constituted by overlapping lineages of apprenticeship.

"We were young and full of piss at that time. You know how you are when you're thirty years old and you think you can do anything. More hard-core as far as historic principles and all that. That's how businesses like this get going. It's part derring-do and part ignorance and part planning and part fate. Luck. You get hooked up with the right people who want to buy your work and pay you to do it."

Some of this luck took the form of a cultural moment. "The early music scene just exploded in the seventies and eighties. All the harpsichord builders were busy, and all the people who made recorders and wooden flutes and all that—it was huge. At this point, the strong have survived, and it's greatly more concentrated, not as diffuse as it was, and the market is not as large. But we're lucky—we're tied to the church organ thing. When one denomination is giving it up, the next one is reviving it. The Catholics now [2007] are going after high-quality music because the pope is interested in that. He's a German and a pianist, and he knows good pipe organs. He doesn't want Catholic churches to be having guitar bands anymore." John pronounced "guitar" with the stress on the first syllable.

I asked if the pope was actually sending word down. "Oh yeah! So for example St. Mary's Cathedral in Austin, Texas, they tore out their pipe organ about twenty years ago and have had an electronic organ, and now they want to buy a new one. I think it's a great thing for the church. There's a renaissance going on. People are conservative, I think. They have a yearning for their roots. They're practicing the Latin mass in this parish. They used to do it in the closet so nobody'd know they were doing it, and now it's okay."

This yearning for roots creates a complex set of demands and opportunities that Taylor and Boody has to be responsive to. "We're in a strange place, because here we are, making a re-creation of an historic thing; we have documentation of organs going back three, four hundred years, and have gone and studied them in Europe.

But we're weird: we're trying to make a living, for one thing, and make this whole thing practical, so we have to make the parts in good order and build to the contract price, which is an *insane* thing to do for what we're producing, and we have to make something that's going to perform technically or our customers are going to be coming back and getting on us. At the same time the ethos of the instrument has to be authentic—that's why people are paying big money for us to do the work. So we're jammed in the cracks, trying to make a realistic business out of this and at the same time be as authentic as we can. The *thread* of what we're doing is totally authentic."

As we talked, John situated his trade in the larger currents of modernization. At some point the making of organs "went one hundred percent industrialized factory building," but the results were inferior. "It didn't satisfy, so there was a retrenchment. There were also electronic organs available. But there's a part of the population that will only take *this*." John swept his arm dramatically to his drafting table, where a freshly penciled front view of his latest commission lay. "It's a totally handcrafted, handmade object, and some people are willing to pay an enormous premium for it. *They realize the performance musically is superior. And that's the only justification.*"

In fact the demand for organs such as Taylor and Boody make was created by an astonishing musical experience—the discovery of a *sound* that had been covered over by sediments of changes in the organ, and had to be recovered through a kind of archaeology that was at once an engineering project and a cultural project, guided by musical considerations.

THE ORGAN WARS

Unlike, say, a piano, the organ assumes an indefinite variety of forms as organ builders try out new ideas; it has always been an instrument in flux. Organs also vary according to national cultures and the liturgical practices of different denominations. Such practices change over time, as do the architectural trends embraced by different communities of faith. The physical space in which an organ is installed determines a good deal of the character of the instrument. Organ builders, then, are part of a conversation among musicians, architects, congregants, and even theologians, which goes back over half a millennium. More than any other musical instrument, as an installed fixture the pipe organ is a situated thing, impossible to understand without reference to its history.

This ongoing conversation about organs can be contentious—in the organ wars that have erupted at various points, nothing less than the fate of men's souls has been at stake. This becomes most clear in traditions that reject the organ altogether. The Puritans who settled in New England rejected pipe organs as a spur to idolatry and pridefulness, the handiwork of Lucifer. One Eastern Orthodox theologian, Pavel Florensky, denounced the organ as an embodiment of Renaissance humanism (that secret rot at the heart of Western Catholicism), producing a sound "too slow, submerged, and alien, too engulfed in the darkness of human nature, for the crystalline transparency of Orthodox liturgical life."[4]

There have been quarrels within the camp of organ lovers as well, and in the twentieth century these have turned on musical considerations more than theological questions. Yet the passions on view in these quarrels suggest it is the larger issue of "modernity" that has been at stake, no less fraught with consequences for men's souls than the theological battles of another era.

The first shot in the twentieth-century organ wars was fired by Albert Schweitzer. The foremost interpreter of Bach on the organ at the turn of the century, Schweitzer had visited the Liederhalle in Stuttgart in 1896 to hear the new organ that the newspapers were raving about. It was played by an organist whom Schweitzer held in high regard. In his autobiography *Out of My Life and Thought*, he writes, "When I heard the harsh tone of the much belauded instrument and in a Bach fugue which Lange played to me perceived a chaos of sounds in which I could not distinguish the separate voices, my foreboding that the modern organ meant in that respect a step not forward but backward suddenly became a certainty."[5] Schweitzer went on to write a pamphlet that essentially inaugurated the organ reform movement.

To understand how something emphatically "modern" might have struck Schweitzer as "a step not forward but backward," it is necessary to know the trajectory the organ was on at the time. It was becoming a gadget, impressive more for its technical ingenuity than its musical qualities. It also got much, much bigger. This process, which was just gaining momentum as Schweitzer wrote, culminated in the 1930s with such exaggerated instruments as the Boardwalk Hall Auditorium organ in Atlantic City, which has over 33,000 pipes (the exact number is unknown), driven by blowers with a total of six hundred horsepower, at about thirty times the wind pressure of a Baroque organ. *The Guinness Book of World Records* calls it the loudest musical instrument in the world, producing an "ear-splitting volume, more than six times the volume of the loudest locomotive whistle." Because of the wind pressures involved, the pipes are strapped down, lest they launch through the roof. The biggest pipe weighs over three thousand pounds and is sixty-four feet in length. It produces a tone that is in fact not a tone; at eight hertz, it is roughly what you would hear if a military transport helicopter happened to be hovering overhead. The complaint, then, is that such an organ is *not very musical.*

"ELECTRICITY IS HERE TO STAY"

The reform inaugurated by Schweitzer was directed, in the first place, against the electropneumatic control that had replaced mechanical keyboard actions. Electropneumatic control made the action easier, allowing higher wind pressures and hence greater volume, with no corresponding increase in effort at the keyboard. Overcoming the necessity of direct mechanical linkages running from keys to pipes, electropneumatic remote control also allowed a proliferation of stops, corresponding to different sounds. The "orchestral" organ was born; it seeks to imitate every instrument of the orchestra. Through technological progress new things were possible; the limitations of the organ, and therefore also its distinct character, had begun to dissolve into open-ended possibility: the organ as synthesizer.

But since an electropneumatic action has its own inherent "time constants," as a physicist would say (here, the time required for a valve to open and shut), the organist has less control over his phrasing; the keyboard has little sense of "touch." Schweitzer argued for a return to mechanical action. Yet the organ builder Lawrence Phelps tells us that what Schweitzer said about mechanical action and ideal phrasing "was easily passed over as too idealistic and out of touch with reality, for everyone knew that electricity was here to stay."

In the musical world as elsewhere, there seems to have been a sense of techno-inevitability, a readiness to regard technology as a force with its own magical imperatives, rather than as an instrument of human intentions. The saying "Electricity is here to stay" suggested that the growing prevalence of electricity was due to the working out of some rational necessity, and to deny this was to reveal oneself as "out of touch with reality." Such a reflex is often part of the makeup of those who take themselves to be the most forward-looking. Yet the progression of an engineering art seems to require a freer sort of relationship to the past without the progressive prejudice, as well as a critical stance toward one's own times. In retrospect, it is the enthusiasts of electricity who appear to have been caught up in a strange idealism, a willful disregard for *function*, which, after all, is the whole point of technology.

Schweitzer's critique of electropneumatic action might be taken as an instance of the wider antimodernist sentiments that were circulating at the turn of the century on both sides of the Atlantic. Such sentiments rested on intuitions of something gone amiss in modern culture. But many found it difficult to back up these intuitions with arguments, so they were dismissed as manifestations of romantic discontent. Yet Schweitzer's critique of the organ of his day was detailed and at bottom mechanical. It was emphatically rational, and pointed to something irrational in the heedless embrace of new possibilities merely because they are possible. Schweitzer prevailed, enough so to create a critical mass of dissidents in the musical world. The defects of the organs he criticized became visible *as* defects because eventually it dawned on people that they were not very musical.

Let's pause here to note that the critique of current trends in automotive design that I offered in Chapter 4 closely parallels Schweitzer's critique of the organs of his day. Both are arguments for direct mechanical linkages and hence greater "touch"; both demand that an instrument be supple in transmitting sensorimotor information. And both critiques require bringing a certain cultural disposition into view: the fetish of automaticity and disconnection. As we have seen, this is the deep tendency of a culture that connects the upward march of human freedom and dignity to an ever greater abstraction from material contingencies.

It is fashionable to scoff at the idea of a "privileged" moment in culture (for example, the Baroque era for organs, or the decades before the 1990s for automobiles) that is better than any other moment. Let it be conceded that the orchestral organs of the early twentieth century must have swelled the worshippers of that time with an aesthetic-religious experience no less real than that of their Baroque predecessors. To speak of decadence, then, smacks of nostalgia, that thought crime that popular writers are quick to detect in anyone who glances backward.

Yet our low regard for nostalgia often seems not to rest on some substantive standard of excellence, in light of which a preference for the past is seen as missing the mark, but rather expresses idolatry of the present. This kind of "forward-thinking" is at bottom an apologetic species of conservatism, as it defers to and celebrates whatever is currently ascendant.

ANCIENTS AND MODERNS

But what is this Baroque sound that the organ reformers loved so much? Does it go beyond the control of phrasing made possible by mechanical action? To get some feel for what the organ reform movement was all about, I decided to call an organist I knew, Frank Archer. Frank agreed to meet me at the First Presbyterian Church in Farmville, Virginia, for an informal organ clinic, on an instrument made by none other than Rudolf von Beckerath, the German teacher of Taylor and Boody's teacher.

Frank sat at the honey-colored oak console in the empty church and played a few single notes. "You hear that? Chiff." I wasn't sure what I was listening for. "It's that breathiness in the initial attack of the note. It's different in different stops, and it's crucial to the sound of a Baroque organ." Frank pulled out a few stops, one at a time, each corresponding to a group of pipes designed to produce a particular sound, and after a couple of minutes I could hear what he was talking about: a faint rush of air, audible a split second before the tone, that gave the note a softly percussive quality. With the chiffier stops, it was like listening to a human singer at close range, a husky alto, and almost feeling the hot puffs of air from her lips against the nape of your neck with each phoneme.

Frank explained that chiff makes overlapping melodic lines distinguishable in contrapuntal music, since the initiation of every note is marked by this breath. Without chiff, the music becomes soft and muddy. As organs are instruments played in large, resonant spaces, keeping separate voices distinguishable is an inherent problem for the same reason that voice intelligibility is a problem in movie theaters: the ear receives sound reflected from various surfaces, following different paths and therefore arriving at the ear at different times. What began as a distinct note becomes a blur. In movie theaters, this problem is controlled by covering the walls and floor with sound-deadening materials that minimize reflection. In an organ-listening space, such measures would kill the powerful resonance that organs have traditionally enjoyed in their stone cathedrals, and that composers assume in writing for the organ. Sitting at the console, Frank related how Beckerath had returned years after the initial installation to check on his organ at Farmville and was horrified to see that cushions had been added to the pews for the comfort of the parishioners, and carpet to the floor. In a thick Hamburg accent, he asked where he might rent a truck. "I vill take zees to ze dump!"

The sustain of an organ, crucial to its aesthetic character, is largely an architectural fact—the church is part of the instrument.[6] Yet this desired resonance is in tension with the listeners' ability to discriminate separate voices. Chiff eases this tension by punctuating each note with a nontonal attack; a sound that will not reflect because it is not "coherent" as a wave form, and decays quickly.

The acoustical logic of the ancient organs' chiff, then, is

impeccable. As a pneumatic instrument, the organ has within itself the resources to overcome a problem inherent to its site, and the Baroque organ builders seem to have understood this. Yet chiff is something that had to be recovered by the organ reform movement, as it had been lost due to the advent of deliberate nicking of the languid (part of the windway of a pipe). Why would one do this? Pursuing this question reveals a fascinating case of changes in musical taste that are tied to a kind of cultural sedimentation, and rest ultimately on a process of physical decay. Writing in 1969, Lawrence Phelps explains:

> Why was nicking introduced anyway in the early 18th century? Was it really because of a changing taste? I think not. I have suggested that the practice of nicking was introduced to make new pipework cohere better with the pipework of previous builders; older pipework that had lost its sizzle due to the aging of the metal and the wearing of the edge of the languids by the passage of a few decades of wind. This aging and the change it produced in the sound of pipes was a perfectly natural physical phenomenon. Making small nicks on the edge of the languid was found to bring about a similar effect artificially in new pipes, but the effect was of course compounded with each generation of builders. Thus, the practice of nicking eventually, though very gradually, brought about the change of taste that produced the smooth, lifeless, opaque tone so common in the flue-work of even the best of the romantic builders and which reached ridiculous extremes in the early decades of this century. We, moderns, having been re-exposed to the natural sound of undoctored new pipes, have found that it restores meaning to the music and the instrument we love so much, and we know that this sound is here to stay.[7]

Through sediments of forgetfulness, the original excellence of the pipe organ fell into oblivion, "compounded with each generation," as Phelps says. The muddy sound characteristic of erosion of the languids became an aesthetic fact of the West, culturally established as the horizon within which organ music grew comfortable. The crisper musicality proper to the organ had to be recovered through an archaeological effort. Digging through the strata of confusion, organ builders discovered the root cause of their discontent with the twentieth-century sound: metal erosion. Ironically, these antiquarians opened the way for real progress; their work led to a *discovery* that made possible an unexpected musical freshness—the sound of "undoctored new pipes."[8] In reverse-engineering the old organs, then, the reformers weren't simply finding technical tricks to accomplish an end they already had in mind. Rather, they were discovering the standard aimed at by those earlier builders. In doing so, they came to affirm that goal as the proper one for "the music and the instrument we love so much."

This entailed not just a technical accomplishment but a reorientation. In a sense, the judgments of the ancients have become their own judgments. To the casual bystander this looks slavish; it looks like what John Stuart Mill called "the ape-like faculty of imitation" and Kant called "the self-imposed immaturity of mankind." But in fact it reflects an earned independence of judgment. The organ reformers' discontent with the present loosened their deference to it, strengthened their opposing critical muscles, and prepared them to be turned in an unexpected direction.

It seems we need to supplement Kierkegaard's psychology. He taught us that reverence is a prerequisite to rebellion. The organ reform movement sheds light on the other side of this coin: a readiness to rebel—against the self-satisfaction of the age—seems to be prerequisite to discovering something you judge worthy of reverence. To affirm something in this way, freely and with discernment, is surely one element of what it means to be an individual.

Notice that in this movement, liberation is the beginning, not the end, of an education into independence. As the term "liberal education" suggests, to be educated requires getting free from—led out from—taken-for-granted certainties. But when we go deep into a practice, so that its ends become our own, we find ourselves situated in the jig that surrounds the practice, for example the rich and contentious inheritance of organ making, which is disciplined by musical considerations. This jig imposes some definite shape on one's own life as one who is devoted to making good organs. Within it there is room for, indeed a necessity of, interpretation of the standards and of how best to realize them, so the organ maker necessarily puts his own stamp on his product. We can understand this as a richer, more highly elaborated version of what the short-order cook does when he improvises to meet the demands of the kitchen.

After I spent the morning with Frank at the console, patiently demonstrating the musical significance of the various sounds made by the organ, he allowed me to go inside the instrument and watch the action while he played some Bach. Clambering around on a catwalk with a flashlight, I felt encased in a breathing organism as Bach's musical intelligence, and Frank's, surrounded me like a heartbeat. The distinct lines of the Toccata and Fugue in D minor were visibly realized as the layered movements of Beckerath's artfully arranged aluminum tracker rods, each with ball joint swivels just like a motorcycle shift linkage, sounding notes above me and below me, to the left and the right in interlocking themes. It was like being in the middle of a lively dinner conversation of close friends, my head swiveling to catch the gestures accompanying each syllable of Frank's interpretation. At the end of the day, I felt better prepared to return to Taylor and Boody and go deeper into their world.

LEATHERING AND NIBBLING, AND DIFFERENT TEMPERAMENTS

As I followed John across the assembly room, we came across Chris Peterson ("Pete") releathering the wedge-shaped bellows for the Henry Erben organ, which would be pumped by the organist himself while playing, using his feet. "This is very nice what you've done here. Did you do this with your router?" John asked.

"I did it with my chisel." Pete had opened up a large crack in the face of the bellows, resulting in a square hole about an inch on each side, then plugged it with a new piece of wood cut to fit precisely. Finer cracks he covered with leather on the inside; the bellows needs to be completely airtight. "I put a Harley sticker in there too, hope you don't mind."

I wasn't sure if he was joking or not. People who make stuff, and people who restore old stuff, often want to leave a message to be found by some bloke in the future. It is fun to discover these, and to contribute. In any case, Pete clearly didn't regard this organ as an inviolable fetish object. It had been entrusted to him for a reason—to make it functional once again—and to do that, he was going to have to make it his own for a spell, before passing it on to the future.

Penciled on the wood of the bellows were the words "Pull open 7 ½." This had apparently been written at some earlier point in the organ's history. Pete said, "What I have down is one hundred fifty-five millimeters." I took him to mean the tangential travel of the bellows at the point farthest from its hinges.

John pulled out his tape measure, with millimeters and inches both, and said, "Seven and a half inches is one hundred ninety, not one hundred fifty-five."

Pete said, "These hinges were in disarray, so that might have something to do with it."

John suggested the hinges were "cheap Victorian cast iron."

Pete allowed that "they drilled like cast" when he restored them, but thought they had been put on when the bellows was releathered in 1957, and that the problem lay with the location of the ribs of the bellows in relation to the hinges, causing excessive wear. Pete explained that he builds a bellows differently to avoid this problem.

I asked how long the new leather is expected to last.

"Long enough that I won't have to do it again."

John explained that the longevity of leather used to be a bigger problem, but it has been alleviated by getting back to a vegetable-based tanning process, which leaves more oil in the leather than harsh chemical tanners.[9] "We've used Cabretta goat leather here, which seems to stand up quite well in movable pieces. It's got a lively feel to it." He handed me a piece of thin white material that was unbelievably supple, like an old T-shirt.

"Pete has used all traditional hide glue as well, for attaching the leather to the wood, and that's so the next guy who has to re-leather it can get it off. You just heat it up with some hot rags; that's how Pete got the old stuff off. With a modern glue, you'd be bleeding trying to get it off."

Pete said, "You'd have to sand the leather off, or hand-plane it off, or cut it off with a chisel."

I asked, what if you just wanted to get the job done and weren't concerned for future restorations? In unison, Pete and John said, "Titebond." (This is a popular brand of wood glue.) In many of the details of organ making, the recovery of traditional techniques seems to be motivated not by a hankering after the past, but rather a concern for those who will come later.

Shannon Regi makes all the wooden pipes, which have a different sound than metal pipes; most organs include both types. She came to Taylor and Boody looking for a short-term job after college, and never left. Currently she was working on the reed pipes for Taylor and Boody's Opus 57, which is inspired by an organ made in 1800 by David Tannenberg. John Boody estimates the organ will take about fourteen thousand man-hours to complete. As the name suggests, reed pipes have a reed that vibrates against a shallot, exciting the air column in the pipe just as in a clarinet. Shannon was fitting resonators to the reed blocks, intently nibbling away at them with very small cuts while standing at her bench against a west-facing window. "She works for eight hours and there's this little tiny pile of chips under her workbench and you wonder what she's doing all day." John's floor boss ribbing was clearly in jest. Did it perhaps carry an edge as well? He is the boss, after all. The slightest curl at her lips, Shannon's gaze didn't shift. She ignored John and kept at her pipes. For they indeed seemed to be *her* pipes, and she was too absorbed to acknowledge our presence. John continued in a vein that seemed to exonerate small chip piles. "We do a lot of work like this where all the parts have to be fit together, and every part is different, and every part has to be paid attention to, measured, filed, and fit together.

"All the parts are different sizes because they're related to the musical scale. We can get efficient with some parts, like tracker rods—we might make a thousand at once. But generally everything is one-off." Not only do many of the parts within a single organ scale with pitch, but one organ may have a different tuning than another, with correspondingly different dimensions throughout. The tuning of a keyboard instrument is called its temperament, and there are several different temperaments extant, corresponding to the kind of music an organ is designed for.

John pointed out that, ironically, his shop is less mass-production-oriented than a sixteenth-century organ shop would have been, since he is not making many copies of a single kind of organ. The explosion of historical knowledge since the 1960s, spanning different regions and epochs, and the corollary demand for diverse kinds of organs, means that there is very little repetition in the shop.

THE BOSS

I asked John about his role as boss. "The clock is ticking here all the time. It takes a hundred thousand a month to run this place—that's our operating cost. That means we've got to turn one point two million a year, and if we're not, the ship is sinking. So the guys may not feel it, but George and I are under that strain, and Cindy, we're talking about that all the time. We try to be very organized." Cindy is the bookkeeper.

What about motivating employees? "We have to encourage them to have good work habits. We post a schedule of work to be done. And we have rules." I asked John to elaborate. "Oh yeah, we have a handbook." He began to quote from the handbook. "Come to work. Work forty hours a week. Come during the core time, which is eight-thirty to four-thirty. Clock out for lunch. Take half an hour for lunch, hopefully not too much more than that. Be diligent." As he pronounced these straightforward imperatives, it became clear that John is innocent of such developments as "liberation management." He's a boss, not a life coach or facilitator of personal growth.

On one of my visits, Robert Hanna was refinishing the casework of the Henry Erben organ. Hanna is not an employee of Taylor and Boody. A specialist in finishes, he is a journeyman in the original, literal sense. He goes wherever the furniture is, traveling by car because the airlines do not allow the chemicals he carries. He is at the very top of his profession, a conservator of multimillion-dollar pieces of furniture, and he makes a lot of money. He is essentially a forensic chemist; he speaks of particular oils, shellacs, acetones, and methylated spirits. He is also a cultural historian, and gave me an impromptu dissertation on the variations in American furniture by regions, periods, the local arboriculture, the ethnicity of the cabinetmaker, and the particular tradition within which he worked (for example, Shaker).

John Boody remained throughout this conversation; he seemed as interested to talk to Hanna as I was. This was a bit of a surprise, given that John was paying Hanna close to a thousand dollars a day, and here we had taken up an hour of his time. John's view of productivity appears to be a complex one, as it evidently includes talking, the kind that frames the work going on in his shop in historical context.

When Hanna criticized certain illustrious cabinetmakers in the pantheon of Americana, John noted the parallels with organ building, where in restoring organs by old masters you sometimes "see where they were disastrous in how they approached something, choosing poor materials or not understanding how to do something so it would last a long time. You learn from their mistakes." Over the next several months it became clear to me that such conversations combining history and engineering, guided by a visceral concern for the excellence of organs, are a regular part of working at Taylor and Boody, however much John might groan sometimes about "people standing around flapping their gums." John himself is the worst offender, instigating these conversations as he moves through the shop.

The conversations seem to enact a particular kind of authority that John has in the shop. It is more like that of a teacher than that of a manager, and I think it is best understood in connection with the idea of "the thread" John mentioned in our first conversation. His role appears to be that of taking what can be learned from the tradition, interrogating it critically, and linking it to an image of perfection to be achieved in the future. The conversations help to make this a shared image. They also locate the goals that the shop is aiming at in the dramatic arc of a history, extending back into the past and forward into the future. In this way, the work is enlivened by a sense of going *further* on a trajectory they have inherited. Their ingenuity is focused both by the shifting contours of organ making over the centuries and by the timeless standards of engineering, specifically as they contribute to musicality. These provide two different sets of criteria for their performance as organ makers, neither of which is simple enough that it could be reduced to an explicit recipe. Each requires what Polanyi called "personal knowledge," just as we saw in the case of scientific apprenticeship. Further, these two criteria of historical coherence and musicality aren't simply in harmony, nor are they simply at odds with each other, but rather exist in a fruitful tension. This tension seems to be the upshot of "the thread." John's role is that of keeping it taut by keeping it in the shop's collective awareness through these conversations.

"THE ORGAN HISTORICAL SOCIETY WILL HAVE NOTHING TO DO WITH ME"

The sense of inheritance that the employees are working within is experienced not as a burden, but as a source of energy. If regarded differently, however, it could easily be a source of complacency, a dragging anchor of stasis. This became clear in a certain controversy related to me by Chris Bono, who has been working at Taylor and Boody since 1988. "The Organ Historical Society will have nothing to do with me," he said.

A musician as well as an apprentice organ maker in 1988, he volunteered to restore the organ at St. Francis, a Catholic church in Staunton where he got a weekend gig as organist. "The organ wasn't playable in 1988. It would make *some* sound, it would kind of wheeze away, but there were pipes leaning every which way, I mean, it was dangerous. Somebody had taken tubes of silicone bathtub caulk and tried to fix the leaks." It was also covered in plaster dust; under that was coal dust. "So I took the whole organ apart. The concept went from cleaning the organ to 'Well, now that I've got it apart, it wouldn't be all that difficult to add a this, to add a that.' I think I had my first child before I was done with the darn thing. But in the end the church got a good organ that *I* enjoy playing."

As the disapproval of the Organ Historical Society suggests, Chris's concern was for musicality more than historical accuracy—"a good organ that I enjoy playing." The historian adopts a neutral or nonjudgmental stance toward the facts of the past. The preservationist's love for the old is similarly nonevaluative: it requires of it nothing other than that it be old. The St. Francis organ represented a certain stage in American organ building, and a case can certainly be made that it ought to have been restored according to the original intent of its makers. But the musician wants a good instrument. As an organ maker as well as a musician, Chris could not adopt a nonjudgmental stance toward the facts, as the historian does, nor a deferential one as the preservationist does. Merely *as* facts, they do not impress him.

Chris made Baroque-style wind-chests and action for the very non-Baroque St. Francis organ. From a certain perspective, this looks perversely anachronistic. But his high regard for the testimony of the more distant past, as against the more proximate past, is informed by the timeless demands of musicality. In striving to meet these demands within the forms available, the organ maker is engaged in a form of rational inquiry. His work wouldn't have this progressive sense—this quality of being a kind of inquiry—if he didn't adopt an evaluative stance toward the facts he has inherited, a way of regarding them that is illuminated by "the good." Here the good in question is simply "a good organ that I enjoy playing."

THE VOICER

The pipes of an organ must be minutely adjusted to achieve a desired sound. In his exhaustive 1905 treatise *The Art of Organ-Building*, George Audsley writes,

> Experience—tedious and expensive—together with individual talent and infinite patience, are the chief factors which combine to form the Voicer. It does not take a great organist to become expert in the art of voicing; but the voicer should have knowledge of the rudimentary principles of music and the laws which govern the production of musical sounds. In my experience, I have found that men with good voices and capable of using them well in singing, good violinists, and men endowed with patience and great mechanical skill, make good voicers.[10]

On my third visit to Taylor and Boody, I mentioned to John Boody that I had traveled to Farmville and spent a day with the Beckerath organ, and this seemed to convince him I was serious in wanting to learn about organs. He pushed open the door to a room I hadn't yet seen. Far from the whine of saws and the pounding of drop hammers, the voicer does his work in the quiet sanctum sanctorum of the organ maker's shop, where pipes are made to speak.

Here John introduced me to Ryan Albashian. I half expected some Gandalf-looking wizard, but Ryan is impressively normal-looking, even athletic, and appears to be in his early thirties. "Ryan is an organist who then trained as an organ builder, and is now the head voicer here."

Ryan pauses for effect and says, "Yeah. Head voicer." The way he says it, it sounds like he intends some double entendre, though I can't be sure. He then brings the pipe he is holding in his hands to his mouth and blows on the end of it, exciting its harmonics. "You hear how badly this . . . [blows again] I mean, this thing is just completely not behaving. Part of the reason is that the cutup is ridiculously low."

John: "Did Jeff look at these to make sure there aren't open seams in the back and such?"

Ryan: "I don't know. I haven't had any problems with that, and the seams look hot on all these pipes." I take him to mean they look like they've got fresh solder on them. Ryan is taking the original pipework of the Henry Erben organ, listening to it, going through the new pipes made by Jeff and Robbie, and fitting them all in to work together musically.

John has other things to attend to, and leaves me alone with Ryan. I ask, "What are you actually doing when you voice a pipe, and what are you listening for?" He plays some notes on his voicing station, which is essentially a compact pipe organ in which the pipes simply rest on felt gaskets on a small wind-chest, directly above the individual keys. There is a round gauge on one corner of the keyboard that registers wind pressure.

"These sound pretty even. I've worked on all these. I'm trying to pick out one that might not sound so good in terms of what we call the color of the sound. It's hard to explain. It's the difference between the car horn on a Ford and the car horn on a Chevy—there's going to be a difference." He zeroes in on one pipe, pressing its key over and over in rapid succession. "Okay, that one's sort of . . . hmm, borderline. I had trouble with that pipe. But it's good: I mean, for something that's made in 1830, I've recovered it to life pretty well. You hear how when it initially speaks it kind of goes, 'yeeaahh, yeeaahh.'" Ryan makes a sound like an adolescent boy's voice breaking, from high to low in the course of a single syllable. Now he presses the key several times through only the initial part of its travel, so the pipe is just beginning to speak, but never getting to full song. He imitates what he hears with a weak falsetto "ha-ha-ha-ha-ha."

I ask him if he's zeroing in on the initial attack of the note.

"The initial attack and the ongoing sound. It's kind of flutey and . . . [Ryan makes a sound like someone discreetly coughing up phlegm in a quiet seminar room] windy."

"Is that what you call chiffy?"

"No, chiffy is when they go 'choong, choong, choong, choong.'" He makes a sound like a steam locomotive leaving the station.

"They have a very definite speech, but it's a very hard kind of speech, with a kind of pop at the beginning. I mean, sometimes you can get away with that in a really big cathedral kind of space, but in most churches you can't." What had taken my ear some time to discern under the guidance of Frank Archer, namely chiff, is for Ryan a noise so overbearing that it does not enter into his considerations with this organ.

Still speaking of this one pipe, he says, "It's a little bit quiet too. But it's acceptable. I'm not trying to change the character of what this organ was. That's the challenge, to make the pipes work but not take away from the integrity that they had." Ryan plays an ascending melodic line, then a descending one. "They're very even, I mean they're really pretty good."

The windways of organ pipes get eroded over time by the passage of air through them, and this can degrade their sound. But there is another respect in which they improve with use. Ryan explained how metal organ pipes change over time due to being played. The maxima and minima of vibration along a pipe's length, corresponding to the fundamental frequency and higher harmonics of the pipe, apparently bring about a molecular change in the metal such that the pipe is more easily excited after a few centuries of use. In some respects, Ryan says, organ pipes "sound their worst at the very beginning of their life." Over time the sound becomes softer, less harsh.

"Part of the challenge of the Yale organ [Taylor and Boody Opus 55] is that they wanted an organ that was historically inspired and sounded like these organs that they heard in Europe. Well, that's all well and fine, but from the voicing standpoint, I had to think to myself, What can I really get away with here, and *what will the results of my actions be four hundred years from now*? If that Yale organ is still in that chapel four hundred years from now, and we assume it will be, then what is going to have happened to the voicing I did? Is the organ going to be so politely smooth that it's got no life in it? So I had to be conscious of that and not overdo anything. Yet we wanted these to behave like old pipes.

"I guarantee you there's no pipe in the Yale organ that looks like this." He points to the Erben pipe. "Not at all. To the naked eye, looking at almost any pipe in the Yale organ, your eye is not drawn to any kind of manipulation that was done in the windway. It's so subtle. By all accounts, we're told it's one of the most successful results that anyone's ever heard."

Ryan clearly means to rival or outdo those previous organ builders whose activities have imparted a shape to his own life, in an ongoing dynamic of reverence and rebellion.

"One of the things I've discovered . . ." Ryan pauses. "I have to be careful not to take credit for this." As a voicer, Ryan evidently feels himself part of a community that has certain unwritten rules. The knowledge of this guild is largely shared—they are all in conversation with one another, whether explicitly or not, and with the past masters. But the knowledge must be put into action, and in doing so there are occasions for the pride in discovery that comes with experimentation.

"To my knowledge no one else has ever suggested that this is the case, and in fact it took me a long time to convince a really, really fine organ builder friend of ours down in Chattanooga. But *I know for a fact* that pulling the upper lip out makes the pipe faster, and pushing the languid down also makes the pipe faster, but *they don't do the same thing.* The upper lip is like the starter, it controls the starting energy of the pipe, how well it starts, how fast it starts. The languid, in here, controls the ongoing sound."

Ryan pauses to concentrate on what he is doing: "cutting up" one of the pipes for the Erben organ. He cuts to a line scribed just above the existing upper lip, which was Jeff's approximation, then bevels the new lip with a knife wielded with a pulling motion, his arm tensed as it would be in any action that requires both high effort and precise control. The soft metal of the pipe, a mixture of lead and tin, gives way to the hardened tool steel of Ryan's knife. He splits the fine scribe mark down the middle.

"The sound quality of the pipe is very much colored by how tall the mouth of the pipe is. The higher you raise the upper lip, the more you alleviate the intensity of the upper harmonics and the more you get the fundamental. But you don't want just fundamental, you want that other stuff too."

Returning to the job at hand, Ryan mutters that Jeff hasn't

properly aligned the foot of the pipe with the body. "Usually he's right on the money. Jeff's a real craftsman. But this is no good. This can make or break the way a pipe plays." The foot is the end of the pipe, and it is cone-shaped. It is soldered onto the main body of the pipe; the mouth is located where the two meet, with the lower lip supplied by the foot and the upper lip by the body. They must be aligned so that the two lips are perfectly parallel.

Ryan shows me the allegedly crooked job, and because I don't want to interrupt his work flow any more than necessary, I pretend to see the out-of-parallel Ryan was exercised about, but in fact I can't. He proceeds to pull the upper lip out on one side, then to use a tiny brass hammer to push it in on the other side. I ask if there is some measuring instrument he might use in such cases, and he says, "Eyeballs."

Tapping away, he announces the pipe is now passable. "A lot of the changes we make are not measurable. I mean, you're talking millionths of a millimeter. It all has to do with what you hear." Ryan takes the pipe and blows on the end of it, producing a succession of different notes. "Okay, the pipe is a little bit fast. You hear it overblows very easily." If you have ever blown on an empty beer bottle, you know that blowing very softly produces the lowest note, and blowing harder produces higher notes.

Ryan puts the pipe to his mouth and plays the fundamental, then the third harmonic (a pipe that is closed at one end has only odd-numbered harmonics), then the fifth, and finally, blowing very hard, the seventh. This whole harmonic series is present when the pipe is played softly, but it is the fundamental that predominates.

Ryan puts the newly doctored pipe on the voicing station and plays it, alternating with another pipe. I think I can hear the overblowing he's talking about: the initial note is higher, before the pipe settles into its lowest mode. Just like a beer bottle.

"Okay, there's a few things going on here. The languid is a little bit low. I'm going to raise it up just a little." Some subtle manipulations follow. "Also the toe hole seems a little large." Ryan commences another operation to collapse the hole at the tapered end of the pipe.

"Voicing is listening. There's very little written material on it. A voicer develops his or her own techniques. You have to understand the science. Friends of mine think it's inherent talent, but not really. The preparation—getting the cutups right, getting them straight, manipulating the windway sizes and the toe holes—you could train a monkey to do that sort of thing. Just give them the list and they do it. If that's done really, really well, that's ninety-seven percent of the work. After that ninety-seven percent is finished, the pipe will make a sound, it'll play. Getting it to play beautifully, that's the last three percent. And that's where I'd like to say the artistic integrity of the voicer is put to use, and it's at that point where the good voicer who has done really well preparing the pipes can take all the math and all the science, if they so desire, and you can start to throw some of that out the window. Now we're not in the science book anymore, we're not studying the math, now we're making music. Now the canvas has been prepared, your base of the canvas has been painted on, you've got the brushes in your hand, and you can feel free to put the strokes down. But you can only do that if the preparation is really well done. If you don't do that, it makes the brushstroke very difficult to get clean on the page."

On Ryan's account, making a musical instrument seems to parallel the process of making music. It doesn't feel very "artistic" to practice scales endlessly; like the preparation of organ pipes, "a monkey could do it." In both cases the expressive element, that last 3 percent, rests on a large base of technical proficiency.

Ryan consults a graph plotting the measured heights of the cutups of the Erben organ's existing pipes. It is a jagged line with a general trend to it. Ryan has drawn a straight line through the jagged one, capturing the trend, and extrapolated it beyond the existing data points to determine the cutups for the replacement pipes. He isn't going to even out the cutups on the existing pipes, as they are part of the character of the organ. "But I guarantee you'd never go into the Yale organ and measure the cutups and find them all over the map like that. Although there *are* little discrepancies, intentional ones, so if anyone ever does get their paws on those pipes, they'll wonder, 'Hmm, I wonder why he did that.'" I ask if these variations were for musical reasons. "No, just having a fun day. Putting my mark on it."

When you are building something that you expect still to be in use hundreds of years later, you apparently imagine yourself as an ancient, being imagined by those moderns to come, the shadowy progenitor of pipes that will be held in hand by some organ restorer of the future. Building his organ in 1830, did Henry Erben form some image in his mind of Ryan, and wonder at the encounter? Like Ryan, Erben did restorations on organs that were old to him, as well as building new ones for the future. Did he pass along some message to Ryan, expressed in the common language of organ makers but inflected by the idiom of his own handiwork? Certainly he did, for every aspect of organ building fits such a description. The narrative of an organ maker's career runs in the larger current of a continuous history. If he becomes an expert maker, his signature is entered into the organ makers' book of common prayer: Schnitger, Flentrop, Beckerath, Noack, Fisk, Brombaugh, Taylor and Boody. This is like the holy book of the Jews, for it is scribbled dense with marginalia by scholars like Ryan, which may or may not become canonical as the tradition of interpretation carries forward.

Ryan puts the pipe back on the voicing station. "That's exceptionally good-sounding. That pipe sounds really good. There's a certain kind of fuzz. There's even more fuzz in this one than there is in C. C is a little bit breathy, and a little bit more flutey. This one is not as breathy, but it's real fuzzy. It's still a little bit fast." He raises the languid by a couple of taps and puts it back on the voicer. "It's a really good-sounding pipe, it's one of the best ones in the whole set yet. Except that it's still a little bit fast." Ryan removes the pipe and taps lightly on the upper lip. "It doesn't need much at all. Again, these are manipulations that you can't really measure." He puts it back on the voicing station and plays it. "That's fine. That pipe is done."

Given how much love Ryan had lavished on a single C-sharp pipe in an organ that doesn't even carry the Taylor and Boody name on the console, I asked John about the culture of his shop. How does he motivate his employees to care about quality?

"They are absolutely the most vicious and greatest proponents of that," he said. "The way to get people most disturbed here is to give them any sense that you're trying to cheapen things, or push them ahead. You think when you start out on something like this that the hardest thing is going to be to teach people to care and value that aesthetic. But it happens, and you cannot convince them to do less. I've never seen them be so angry as when they think that you're trying to cheapen the product, or do it faster—you know, lower the quality."

"So they develop a sense of honor as organ builders?"

"They're impossible. They're totally impossible."

THE DIALECTIC WITH TRADITION

In a kind of mechanical forensic archaeology, George Taylor, John Boody, and a small cadre of other builders with overlapping apprenticeships reverse-engineered the ancient organs to recover the construction techniques and preferred materials of the old masters. John pointed out that centuries ago, as now, the cleaning, maintenance, and tuning of previously installed instruments provided the occasion for an organ "service tech" (as we might call him) to study an instrument. Generally such techs were themselves organ builders, so the activity of reverse-engineering another maker's organ to learn new techniques is itself part of the tradition of organ making. (It is much like the history of philosophy in this regard.) The explosion of information in recent decades, including design drawings and specifications for old organs, John finds "both helpful and distracting"—helpful for obvious reasons, and distracting because "at some point you've got to get on with it." Get on with building your own organs, that is. In doing so, Taylor and Boody seem to adopt a neutral stance toward both history and technology, fraught neither with romantic resentment toward change nor with the kind of uncritical enthusiasm for "high tech" that embraces change merely for the sake of change. Their purpose is to build the best organs they can.

But what is best in an organ? Unlike a space shuttle, the pipe organ is a species that comes to us through cultural traditions, and serves aesthetic purposes that would be unintelligible without reference to those traditions. To start with a "clean sheet of paper" would be to miss the whole point, because to a large extent the history of organs constitutes what it means to be an organ. John Boody and his coworkers are constantly making improvements, and their inventiveness is both limited and energized by tradition—an unusual combination of the spirit of technology with the spirit of loving antiquarianism. These are ethical dispositions, really—the one gets enlivened by new challenges to be overcome, the other finds its dignity in the continuation of old ways. If these seem incompatible, it may be because we moderns have inherited a view that pits the technical spirit *versus* tradition.[11] Partisans of the first will say it embodies reason, and that the latter amounts to little more than inherited prejudice. For their part, partisans of tradition often see in technology a spirit of vandalism that can only destroy meaningful human activity.

But to be in conversation with a tradition *is* a kind of rationality; a mode of thinking that helps us get at the truth about things. This was brought home to me by a story John told about restoring the organ at the chapel of the University of Richmond. This organ was built in 1965 and boasted certain "space age" materials. Thirty-five years later, these plastic action parts and synthetic gasketing materials had melted, and the organ was unusable. Taylor and Boody tore the organ down and replaced the space age stuff with traditional materials: wood, felt, and leather. If the standards of technology are those of functionality, then in this case wood and leather turned out to be higher-tech than plastic. As woodworkers have known for centuries, the dimensional changes of wood with humidity (there is little with temperature) can be accommodated by orienting the grain lengthwise to the dimension that needs constant tolerance. Wood and leather are easily worked with hand tools, have excellent "toughness," and their durability is a known quantity. They are general-purpose materials readily available, rather than proprietary ones tied to the fortunes of one company. With respect to future generations, they make repair work a transparent matter—leather can be sewn and wood can be glued with common glues, whereas the chemistry of plastic polymers is an opaque matter that makes bonding uncertain. Here is a case where space age materials were a bad idea precisely because, ironically, they were insufficiently future-oriented.

Yet those who work within a craft tradition cannot dogmatically identify the good with the old: a living tradition does not consist of a set of static truths passed down. For example, Taylor and Boody will soon undertake the restoration of an instrument in Pittsburgh. "All the synthetic materials are going to go, all new squares in the action, all new trackers. We'll replace them with either wood or carbon fiber." When I expressed some surprise at this last item, John's eyes lit up. "Carbon fiber turns out to be *excellent* material for trackers. It's stable, extremely strong, and stays absolutely straight."

The tradition of organ building evidently consists of an ongoing quarrel about how best to realize certain functional ends. John Boody is engaged in a living conversation, concretely expressed in action, with every organ maker whose work he has examined, and with the authors of every musty organ-making treatise extant. Given the opportunity to examine his organs, one imagines these predecessors would recognize in Taylor and Boody a competent conversational partner, which is different from someone who simply parrots your words back to you. The conversation has a point, and moves along. To participate, an interlocutor must have good manners: he must listen well, contribute with tact, and have that sense of shame that helps you recognize when you have been refuted.

There are external facts that keep the conversation moving along, as this conversation not only answers to the internal goods of organ making but also issues in a product that must please others who are not directly involved in organ making, and may be oblivious to its finer distinctions: the congregation who will use the organ, and pay for it. Thus Taylor and Boody's Opus 64, for St. Michael's Episcopal Church, in Bon Air, Virginia, will have electric stop action. It will still have mechanical key action, which determines the touch of the instrument, but the organist will have the convenience of being able to choose preset stop combinations, which is useful for church services. John says, "We don't think it's essential, but if it makes a difference in whether someone buys an organ or not, we have to think about that." If organ making is to be a livelihood rather than a hobby, then it has to defer to the institutions that pay the bills, and these usually value the music that will be played (often in the context of a fast-moving Sunday service) more than the particulars of the instrument. In this way the organ maker becomes responsible to a wider community of his contemporaries, outside the charmed circle of his guild. He must put his cherished endeavor in its proper place and become public-spirited, out of financial necessity. This helps him gain perspective on his own preoccupations, which otherwise may threaten to become obsessions. As we explored in the case of the motorcycle mechanic, there is a process of triangulation with other people that is built into the organ maker's livelihood; he has to try to square the internal standards of organ making with the wider field of social meaning that we call economics.

As we have seen, the dialectic between tradition and innovation allows the organ maker to understand his own inventiveness as a *going further* in a trajectory he has inherited. This is very different from the modern concept of creativity, which seems to be a crypto-theological concept: creation ex nihilo. For us the self plays the role of God, and every eruption of creativity is understood to be like a miniature Big Bang, coming out of nowhere. This way of understanding inventiveness cannot connect us to others, or to the past. It also falsifies the experience to which we give the name "creativity" by conceiving it to be something irrational, incommunicable, unteachable.

The "going further" that happens in an established tradition, what John called the thread, may be illustrated by a seemingly narrow technical point. John says many of the north German organs of the Baroque period suffered corrosion of their metal pipes from within, and that tannic acid has been identified as the culprit. The tannic acid comes from the oak used throughout those organs, including the wind-chests. Especially in damp conditions near the sea, the air picks up tannic acid from the oak on its way to the lead/tin pipes. Yet oak is the traditional material for Baroque organs. Is there a reason for this? The reason they used oak, it turns out, is that in that part of Europe, woodworms are a big problem; the worms will eat through any softer wood. But in America woodworms are not a problem. So Taylor and Boody are currently transitioning to the use of poplar and pine in their windchests, out of concern for the longevity of their pipes. They haven't observed any problem of corrosion in their pipes, but that is based on observations spanning only forty years. They intend their organs to be in service four hundred years from now.

Taylor and Boody approach tradition with their critical faculties intact. Because the organ needs to work, this puts them on the alert for possibilities opened up by their own circumstances (here, the absence of woodworms). Perhaps it is more generally true that in order to learn from tradition, one has to be able to push against it, and not be bowed by a surfeit of reverence. The point isn't to replicate the *conclusions* of tradition (here, the use of oak), but rather to enter into the same *problems* as the ancients and make them one's own. That is how a tradition remains alive.

The study of the past seems to quicken the activity of the workers at Taylor and Boody rather than burdening it. The history of their craft is constantly being metabolized, absorbed into their own sinews, where it becomes nourishment for vigorous and youthful deeds. In his essay "On the Advantage and Disadvantage of History for Life," Friedrich Nietzsche writes that the thought of being a late arrival in history, though frequently distressing, may, "when grandly conceived, . . . vouchsafe great effects and a hopeful desire for the future . . ." This is so when we conceive ourselves to be "the heirs and successors of classical and astonishing powers, and see in that our honor and incentive." He goes on:

> That the great moments in the struggle of individuals form a chain, that in them the high points of humanity are linked throughout millennia, that what is highest in such a moment of the distant past be for me still alive, bright, and great—this is the fundamental thought of the faith in humanity.

According to the Enlightenment concept of knowledge we explored in "A Brief History of Freedom," the exemplary sort of knower is a solitary figure, and his knowing happens always in the present tense. He is not encumbered by the past, nor does he recognize the kind of authority that operates in communities. His arguments are demonstrative rather than conversationlike in the way we have seen at Taylor and Boody; they float free of any particular historical circumstances or set of lived experiences. Tradition is thus disqualified as a guide to practice. Tradition may convey some truths, it will be conceded, but to be ratified as such the truths in question must be scrutinized by a mode of reasoning that is independent of what came before. To be rational is to think for oneself. For the most part, this Enlightenment understanding views tradition as a darkness that grips men's minds and a habit of inflexibility to be rooted out.

But this view gets a lot wrong. As we saw also in the case of scientific apprenticeship, in the development of any real competence we don't judge everything for ourselves, starting from scratch each morning. Rather, we have to begin by taking a lot on faith, submitting to the authority of our teachers, who learned from their teachers. The individualist conceit that we do otherwise, and the corresponding discredit that falls on tradition, makes people feel isolated.

As we learned from Tocqueville, this isolation brings with it a certain anxiety, which we try to relieve by looking around to see what others—our contemporaries—are thinking and feeling. The rugged individualist becomes the statistical self. The statistical self is the kind that is knowable in bulk, a suitable subject around which to design manufactured experiences. We increasingly encounter the world through representations, produced according to the economies of scale of mass culture. In the worst cases, such as machine gambling, they are guided by a design intention that is inimical to our aspiration to autonomy, even while relying on that aspiration as a psychic hook: manufactured experiences promise to save us from confrontations with a world that resists our will.

The workers at Taylor and Boody are not isolated in this way.

They understand the long story of organ making as their own, and find for themselves a place in it. In this highly situated self-understanding, the excellence they reach for in their work expresses their individuality: an earned independence of judgment, a deepened understanding that is the fruit of their own labors.

Some critics will say that these craftspeople have "retreated from the modern world." I think nearly the opposite. We have come to accept a condition of retreat from the world as normal. The point of the organ shop example is to help us see what it would look like to inhabit an ecology of attention that puts one squarely *in* the world.

EPILOGUE: RECLAIMING THE REAL

The story of the organ shop may inspire some young people to seek out similar ecological niches for themselves, and that is well and good. But most of us find ourselves midway through a life that already has established contours, subject to financial necessities and family ties that compel us to hunker down in whatever spot in the culture and economy seems to be our lot. For us, the utility of the preceding investigation is that it gives us a point of orientation that can help us make sense of the lives we have. Let us revisit some features of our common life.

The problem we began with a few hundred pages ago was that of distraction, which is usually discussed as a problem of technology. I suggested we view the problem as more fundamentally one of political economy: in a culture saturated with technologies for appropriating our attention, our interior mental lives are laid bare as a resource to be harvested by others. Viewing it this way shifts our gaze from the technology itself to the intention that guides its design and its dissemination into every area of life.

The positive attractions of our attentional environment, the ones we willingly invite into our lives, were no less troubling than the unwanted intrusions: we considered the advent of hyperpalatable mental stimuli, and this raised the question of whether the ascetic spirit required for education had a chance. The content of our education forms us, through the application of cultivated powers of concentration to studies that aren't immediately gratifying. We therefore had to wonder whether the diversity of human possibilities was being collapsed into a mental monoculture—one that can more easily be harvested by mechanized means.

I related my experience at a roadside emporium on an Indian reservation and wondered if it offered a glimpse at the end point toward which we are progressing. But in the course of our investigations, a crucial difference between our situation and the historical experience of Native Americans has become apparent. They were subject to an invading foreign power, and rightly understood themselves to be at war for the survival of their way of life. (They lost.) *Our* troubles are native to the regime that we cherish as our own, the product of our greatest virtues as children of the Enlightenment.

The positive examples of well-ordered ecologies of attention that I have elaborated, such as that of the short-order cook, the hockey player, the motorcycle racer, the jazz musician, the glassblower, and the organ maker, have been selected and interpreted for their critical force; they bring out facets of experience that don't fit easily into our Enlightenment framework.

Let me say something about the mode of thinking that I have been attempting in these pages. I take myself to have been doing political philosophy. What I mean is that, like the early modern thinkers I have criticized, I have been doing philosophy in a political, which is to say polemical, mode, in response to a keenly felt irritant peculiar to a historical moment. Centuries ago, the irritant was established cultural authorities that shackled the mind in "self-imposed immaturity," as Kant said. But our emergence from immaturity seems to have stalled at an adolescent stage, like a hippie who hasn't aged very well. The irritants that stand out now are the self-delusions that have sprouted up around a project of liberation that has gone to seed, ushering in a "culture of performance" that makes us depressed.

I am sure that to some people, the reactive motivation of this book disqualifies the effort as philosophy. The temptation in doing polemics is to offer partial truths to counter other partial truths. I have no doubt that the preceding account is partial, and probably in ways that aren't fully apparent to me. But I also take the political mode of philosophizing to be indispensable to philosophy proper, and this is worth dwelling on for a moment.

Philosophy is, among other things, an attempt to understand one's own experience. It therefore has some kinship with the idea of "common sense." But common sense sometimes has to be defended by elaborate arguments directed against other arguments that *cover over* lived experience. Recall Albert Schweitzer's critique of the organs of his time. The muddiness of the sound wasn't directly accessible to experience; the newspapers raved about the new organ in Stuttgart. Schweitzer had to make arguments to uncover the muddiness and to begin the process of reverse-engineering a more musical possibility.

The covering over of experience often began as theoretical doctrines—sediments in the history of philosophy that were first articulated in some argumentative context. If they prevailed, they trickled down and settled as articles of faith, or cultural reflexes. One must deploy sharp implements to clear these away and recover a more immediate intelligibility to life. Philosophizing politically is not something you do only *after* you have figured things out, like Plato's philosopher returning to the cave. It is how you figure things out to begin with.

AN OVERVIEW OF THE ARGUMENT

In the course of thinking about attention, we found that we had to reconsider the boundaries of the self through the idea of cognitive extension. As embodied beings who use tools and prosthetics, the world shows up for us through its affordances; it is a world that we act in, not merely observe. And this means that when we acquire new skills, we come to see the world differently.

We also considered how other people set up shop in our consciousness at a fundamental level, conditioning how we perceive even simple objects like chopsticks and uniformly colored walls. To a significant extent, we know things by having heard about them; they appear to us through the lens of social norms.

This suggested that our cognitive apparatus is thoroughly conventional. We were therefore compelled to ask, what about individuality? How is it possible? For it does seem to be the case that people differ and, if all goes well, a person may develop a deepened understanding that is the fruit of his or her particular biography.

We discovered that individual*ism*, as a doctrine about how we acquire knowledge, arose in a certain political context, that of the Enlightenment, with a polemical intent to liberate us from authority. But the radical self-responsibility that the enlighteners offered as the basis for knowledge seemed to be incompatible with what we had learned about the social nature of knowledge. Therefore we had to ask, is it possible to understand individuality differently? To place it on a different footing from individualism? With the help of Hegel, I suggested that it is by bumping up against other people, in conflict and cooperation, that we acquire a sharpened picture of the world and of ourselves, and can begin to achieve an earned independence of judgment.

But for this to work, they have to be concrete others against whom we differentiate ourselves, not "representatives" of something general, an abstract Public. Yet such abstraction seems to be the tendency of a mass democratic society predicated on liberation from authority. Kierkegaard taught us that rebellion—the moment of individuation—is impossible without a prior reverence. He argued that a flattened human landscape, in which we are embarrassed by the idea of superiority, makes rebellion impossible.

The Enlightenment project for self-responsibility appears to be self-undermining, and has issued in an ideal self that views itself from a third-person perspective. This helped us understand the appeal of early social surveys like the Kinsey Reports (which assigned us to boxes and made it a point of sophistication to view oneself as a representative) and the current appeal of the "wiki" mentality, in which we aspire not to an earned independence of judgment, but rather to participate in the "wisdom of the crowd." Through a logic that Tocqueville laid out, the sovereign individual becomes the statistical self.

Finally, we considered how tradition—a robust cultural jig—

can foster a community of practice in which real independence does seem to become possible (though it is surely never guaranteed). At the root of this possibility is an untimely fact about education that we recovered along the way: it begins by submission to the authority of teachers, as in scientific apprenticeship and craft traditions like organ making.

Obviously, we have not remained narrowly focused on the topic of attention. To make sense of our current crisis of attention required a wider inquiry into the cultural forces and self-understandings that have produced it. In the course of our travels we discovered a few things that bear on that original concern with distraction, and let us now gather these together.

One thing we learned is that the Enlightenment legacy of autonomy talk, persisting as a cultural reflex, can neutralize our critical response to various ways in which our attention gets manipulated. This became most clear in the case of machine gambling, where we found that the gambling industry and its apologists rely on a notion of the sovereign individual to forestall criticism and regulation, even while pursuing "addiction by design" as a social engineering project.

This political thread of argument appeared also in our discussion of the nudge. Libertarian objections to being nudged by the government rest on a notion of autonomy wherein a person's preferences express an authentic core of the self that is not to be tampered with. But this view is hard to sustain, because in fact our preferences are highly influenced by our environment—as sculpted by various "choice architects" who channel our attention for their own interests. For the libertarian to adopt a resolutely individualistic view of the self is to miss this massive fact, and fail in his stated concern for defending liberty.

I sketched our need for the concept of an attentional commons: a concern for justice in the sharing of our private yet public resource of attention. How does the concept of joint attention, which we encountered in developmental psychology, bear on this? Joint attention occurs among two or more people who are engaged in a common enterprise (such as a child playing with blocks with a caregiver), or at any rate in some shared pragmatic situation in which they are mentally present and aware of one another's being present. Joint attention has a natural scale to it, which may have something to do with the limits of physical copresence. It is something that arises organically. Sometimes it is serendipitous, as when two strangers come upon a man passed out in the street and must decide what to do. Or one may deliberately give oneself over to an episode of joint attention, such as a concert, where the whole point is for people to come together and pay attention to something worthwhile. In neither case does one find oneself the object of an engineered effort to appropriate one's attention merely because one's presence in some shared space makes this possible.

Joint attention is an actual experience that we have. By contrast, the attentional commons is best understood as a purely negative principle, by analogy with the "precautionary principle" invoked by environmentalists. The point of being aware of the attentional commons is not to make it happen but to refrain from damaging it; to be aware of the valuable *absence* that creates space for private reverie, and indeed for the possibility of those episodes of joint attention that arise spontaneously and make cities feel full of promise for real human contact.

What this boils down to: Please don't install speakers in every single corner of a shopping mall, even its outdoor spaces. Please don't fill up every moment between innings in a lazy college baseball game with thundering excitement. Please give me a way to turn off the monitor in the backseat of a taxi. Please let there be one corner of the bar where the flickering delivery system for Bud Lite commercials is deemed unnecessary, because I am already at the bar. The attentional commons is an idea that I hope will catch on among those who are in a position to make such sanctuaries happen: building managers, commercial real estate developers, and interior designers. Here is a modest proposal: Could the Muzak be made opt-in rather than opt-out? Once every twenty minutes, somebody *in the room* would have to deliberately hit a button to restart it, and thereby actively affirm "Yes! We want some emo in here!"

We learned encouraging things too. Joy is the feeling of one's powers increasing. The experience of hitting one's flow as a cook, or feeling one's awareness run out through the contact patches of a motorcycle's tires, seems to reveal something deep about the situated, embodied character of impressive human performance. Merely to highlight such experiences provides a point of orientation that can help us assess the manufactured experiences we are offered under "affective capitalism."

Finally, our investigation of "the erotics of attention" yielded some insight into how one might go about escaping the lonely hell described by David Foster Wallace, in which other people are simply impediments to my will. Contrary to Wallace's own take on the matter, I suggested that it is not by freely "constructing meaning" according to my psychic need and projecting generous imaginings onto others that I escape my self-enclosure. It is by acquiring new objects of attention, which is to say, real objects of love that provide a source of energy. As against the need to transform the world into something ideal, the erotic nature of attention suggests we can orient ourselves by a selective affection for the world as it is, and join ourselves to it.

RECLAIMING THE REAL

Affection for the world as it is: this could be taken as the motto for a this-worldly ethics. How much of a departure that would be from the ethics that informs our society we have seen in our consideration of children's television, where the business of forming souls is carried on. In the *Mickey Mouse Clubhouse*, as in many other manifestations of contemporary culture, dealing with reality through a screen of representation serves to make the world innocuous to a fragile ego and the self more pliable to the choice architecture presented by whatever functionary of psychological adjustment is in charge. The world in which we acquire skill as embodied agents is precisely that world in which we are subject to the "negative affordances" of material reality. As I said before, to pursue the fantasy of escaping heteronomy through abstraction is to give up on skill, and therefore to give up on the possibility of real agency.

To reclaim the real would be to go in the other direction. There are, in addition to the slapstick, moments of real physical grace in the original Disney cartoons. When Donald Duck is skating on a frozen lake, he recruits overhanging branches and snowdrifts, incorporating their affordances into a balletic performance in which he is able to do amazing things. His skating is amazing, but not fantastical or magical; it is a heightened version of what you admire when you watch a real skater, or a masterful juggler who is able to improvise and incorporate odd objects supplied by the audience into his act, finding their affordances of weight, rotational balance, and graspability on the fly. To watch the original Donald Duck skate on a lake is likewise to be surprised and delighted by the real. At such moments, the possibilities for beautiful human action in the world as it is—the undiscovered possibilities of *fit*—seem inexhaustible.

This can inspire wonder and gratitude: the most creditable of religious intuitions is available within a this-worldly ethics of attention. For there does seem to be something benevolent in the disposition of things, relative to us. Such are the rules of gravity and buoyancy that *surfing* is possible. That's the kind of universe we inhabit. Being alert to such possibilities, and giving their occurrence in the world their due in wonder: to encounter things in this way is basically erotic, in the sense that we are drawn out of ourselves toward beauty.

DEMOCRACY WITHOUT FLATTENING

What about encountering other people (as opposed to things)? Here too an ethics of attention would be erotic. We are attracted to examples of human excellence, as in our love of sports. Sports seem to be the one realm where we remain unembarrassed by superiority; here the flattening of the human landscape described by Kierkegaard has not taken hold.

In the flattening, we come to view ourselves and others in the third person, as representatives of a generic category. Taken to its conclusion, this egalitarian logic demands that we regard everyone as a representative of Kant's all-inclusive genus "rational beings," so as to guard against special pleading and guarantee moral purity—the kind that requires universality and (says Kant) abstraction from all particulars. Is Kierkegaard's thought—that human difference appears in the form of hierarchy—an antidemocratic thought?

The physical prowess we admire in sports is hard to miss; it requires no special effort of attention. But, as I concluded also in *Shop Class as Soulcraft* (in a section entitled "Solidarity and the Aristocratic Ethos"), our attraction to excellence—our being on the lookout for the choicer manifestations—may lead us to attend to human practices searchingly, and to find superiority in unfamiliar places. For example, in the embodied cognitive finesse of the short-order cook, or the intense intellectual labor that may be required in work that is dirty, such as that of the mechanic when he is diagnosing a problem. With such discoveries we extend our moral imagination to people who are conventionally beneath serious regard and find them admirable. Not because we heed a moral demand such as the egalitarian lays upon us, but because we actually *see* something admirable. Our openness to superiority can connect us to others in a genuine way, without a screen of egalitarian abstraction.

This is not antidemocratic. When the humanity of others who were previously invisible becomes apparent to us for the first time, I think it is because we have noticed something particular in them. By contrast, egalitarian empathy, projected from afar and without discrimination, is more principled than attentive. It is content to posit rather than to see the humanity of its beneficiaries. But the one who is on the receiving end of such empathy wants something more than to be recognized generically. He wants to be seen as an individual, and recognized as worthy on the same grounds on which he has *striven* to be worthy, indeed superior, by cultivating some particular excellence or skill. We all strive for distinction, and I believe that to honor another person is to honor this aspiring core of him. I can do this by allowing myself to respond in kind, and experience the concrete difference between him and me. This may call for silent deference on my part, as opposed to chummy liberal solicitude.

In other words, when a mechanic diagnoses that intermittent electrical gremlin that has been bedeviling your Mercedes for months, be quiet when you write that check, because you are in the presence of genius.

Far from threatening our democratic commitments, attention to rank—the well-earned kind—can put those democratic commitments on a more real foundation.

ON BEING LED OUT, REDUX

To reclaim the real, both in the way we encounter other people and in the way we encounter things, would have implications for education. They are crystallized in the following quote from Doug Stowe, a woodshop teacher and first-class thinker about education: "In schools, we create artificial learning environments for our children that they know to be contrived and undeserving of their full attention. Without the opportunity to learn through their hands, the world remains abstract, and distant, and the passions for learning will not be engaged."[1]

I don't think this is true of every student, but it is true of enough students that we ought to worry about it. A due regard for the diversity of human excellence would include a due deference to the diversity of learning styles. But let us go further: to encounter things directly is more fundamental than doing so through representations, so maybe we needn't regard hands-on education as second-class, and those who require it to flourish as second-rate. Very few of us are scholars by nature, and it seems strange that sitting at desks, looking at books, would become the norm of universal education. Stowe puts his finger on the problem when he suggests that many students are sitting there in class with the silent conviction that what is on offer is "undeserving of their full attention" and engagement. This problem is surely exacerbated by the availability of hyperpalatable mental stimuli. But I believe the more basic issue is the disembodied nature of the curriculum, which divorces the articulate content of knowledge from the pragmatic setting in which its value becomes apparent. By contrast, suppose a student is building a tube frame chassis for a race car. Suddenly trigonometry is very interesting indeed. To reclaim the real in education would be to understand that one is educating a person who is situated in the world and orients to it through a set of human concerns. This is more effective than addressing oneself to a generic "rational being" and expecting him or her to get excited. Our current regime of education has been flattened in this way.

As Stowe's use of the word "undeserving" suggests, at the heart of education is the fact that we are evaluative beings. Our rational capacities are intimately tied into our emotional equipment of admiration and contempt, those evaluative responses that are inadmissible under the flattening. A young boy, let us say, admires the skill and courage of race car drivers. This kind of human greatness may not be available to him realistically, but is perfectly intelligible to him. If he learns trigonometry, he can put himself in the service of it, for example by becoming a fabricator in the world of motor sports. He can at least imagine such a future for himself, and this is what keeps him going to school. At some point, the pleasures of pure mathematics may begin to make themselves felt and give his life a different shape. Or not. He may instead become enthralled with the beauty of a well-laid weld bead on a perfectly coped tubing joint—like a stack of shiny dimes that has fallen over and draped itself around a curve—and devote himself to this art. There are websites for "weld porn," and the mere fact that this is so should be of urgent interest to educators. Education requires a certain capacity for asceticism, but more fundamentally it is erotic. Only beautiful things lead us out to join the world beyond our heads.

NOTES

INTRODUCTION: ATTENTION AS A CULTURAL PROBLEM

1. See Megan Garber, "Advertising That's Squirted into Your Nose on Your Morning Commute," *The Atlantic* Cities blog, July 27, 2012; and Nate Berg, "Bus Riders and Invasive Advertising," ibid., July 30, 2012.
2. This point has been made very nicely by Damon Young in his book *Distraction: A Philosopher's Guide to Being Free* (Melbourne, Australia: Melbourne University Publishing, 2008).
3. *The Onion*, May 30, 2013.
4. This is least true of France, I believe. In the Anglo-American universe, the French are lampooned for their regulatory zeal. But they have a robust sense of the common good, and are sensitive to the small but important ways in which the fabric of everyday life can be degraded if they are not vigilant in defending it.
5. A similar argument has been made by Jaron Lanier in his book *Who Owns the Future?* (New York: Simon & Schuster, 2013). He argues that digital networks that make information appear to be "free" have had the effect of making it harder for people to be compensated for their talents. We become laborers who cheerfully contribute to the value of the network (consider the staggering array of talent on display on YouTube), but that value accrues to whoever owns the network. Our desire for recognition from other people makes us post our best efforts online, and it is the ideologists of "free" who become billionaires while promoting the spirit of sharing.
6. Iris Murdoch, "On God and Good," in *Existentialists and Mystics: Writings on Philosophy and Literature* (New York: Penguin, 1999), 347.

7. "When driving conditions and time on task were controlled for, the impairments associated with using a cell phone while driving can be as profound as those associated with driving while drunk." This is the conclusion of D. L. Strayer, F. A. Drews, and D. J. Crouch in "A Comparison of the Cell Phone Driver and the Drunk Driver," *Human Factors* 48, no. 2 (Summer 2006): 381–91.
8. Ira E. Hyman, Jr., S. Matthew Boss, Breanne M. Wise, Kira E. McKenzie, and Jenna M. Caggiano, "Did You See the Unicycling Clown? Inattentional Blindness While Walking and Talking on a Cell Phone," *Applied Cognitive Psychology* 24, no. 5 (July 2010): 597–607.
9. Frank A. Drews, Monisha Pasupathi, and David L. Strayer, "Passenger and Cell Phone Conversations in Simulated Driving," *Journal of Experimental Psychology: Applied* 14, no. 4 (2008).
10. Walter Mischel and E. B. Ebbesen, "Attention in Delay of Gratification," *Journal of Personality and Social Psychology* 16 (October 1970).
11. See *The End of Overeating* (New York: Rodale Press, 2009), by David Kessler, former commissioner of the U.S. Food and Drug Administration.
12. Arthur M. Glenberg, "What Memory Is For," *Behavioral and Brain Sciences* 20 (1997): 10.
13. Merlin W. Donald, "Human Cognitive Evolution: What We Were, What We Are Becoming," *Social Research* 60 (1993): 143–70.
14. See my article "The Limits of Neuro-Talk," in *Scientific and Philosophical Perspectives in Neuroethics*, ed. James Giordano and Bert Gordijn (New York: Cambridge University Press, 2010), originally published in *The New Atlantis*, Winter 2008, and still available at www.thenewatlantis.com/publications/the-limits-of-neuro-talk.

1. THE JIG, THE NUDGE, AND LOCAL ECOLOGY

1. David Kirsh, "The Intelligent Use of Space," *Artificial Intelligence* 73, nos. 1–2 (1995): 35–36.
2. Andy Clark, *Being There: Putting Brain, Body, and World Together Again* (Cambridge, MA: Bradford Books, 1997), 180.
3. Nelson Cowan, "The Magical Mystery Four: How Is Working Memory Capacity Limited, and Why?" *Current Directions in Psychological Science* 19, no. 1 (February 2010): 51–57.
4. This has far-reaching anthropological implications, which Andy Clark and Michael Wheeler explore in "Culture, Embodiment and Genes: Unraveling the Tri-

ple Helix," *Philosophical Transactions of the Royal Society B: Biological Sciences* 363, no. 1509 (2008): 3563–75. On the evolutionary time scale, we have inherited certain genetic endowments and limitations, but these are massively underdetermining of the resources that individuals bring to the adaptive problems they face. Culture—the particulars of our inherited linguistic, social, and material equipment—establishes the setting for childhood development and all subsequent learning. In the course of that learning our brains undergo both fine-grained and structural changes that are hugely consequential: changes that depend on our *experiences*. There are, then, three time scales that matter for the question of how we come to be what we are: Darwinian evolution, the history of a civilization, and the life course of an individual. This is perhaps obvious, once stated. But it puts limits, which would seem to be fatal, on the explanatory power of evolutionary psychology—that is, on the attempt to explain human behavior as the product of adaptive pressures we faced on the savannahs in the Pleistocene epoch.

5. Richard H. Thaler and Cass R. Sunstein, *Nudge: Improving Decisions About Health, Wealth, and Happiness* (New Haven: Yale University Press, 2008).
6. And, who knows, maybe this is to be preferred. The Protestant is a somewhat cramped human type. One might prefer to spend the evening with someone nudged into saving money (enough so he can pay for the meal), but who doesn't have the deeply internalized ethic of thrift, which easily shades into miserliness.
7. This point has been made by Joseph Heath and Joel Anderson in their article "Procrastination and the Extended Will," in Chrisoula Andreou and Mark White (eds.), *The Thief of Time: Philosophical Essays on Procrastination* (New York: Oxford University Press, 2010).

2. EMBODIED PERCEPTION

1. See, for example, P. Bach-y-Rita and S. W. Kercel, "Sensory Substitution and the Human-Machine Interface," *Trends in Cognitive Sciences* 7, no. 12 (2003).
2. My treatment of the tree is indebted to Søren Overgaard, *Husserl and Heidegger on Being in the World* (New York: Springer, 2004).
3. Maurice Merleau-Ponty writes, "My field of perception is constantly filled with a play of colours, noises and fleeting tactile sensations which I cannot relate precisely to the context of my clearly perceived world, yet which I nonetheless immediately 'place' in the world, without ever confusing them with my daydreams . . . If the reality of my perception were based solely on the intrinsic coherence of 'representations,' it ought to be forever hesitant . . . , being wrapped up in my

conjectures on probabilities. I ought to be ceaselessly taking apart misleading syntheses, and reinstating in reality stray phenomena which I had excluded in the first place. But this does not happen." *Phenomenology of Perception*, trans. Colin Smith (New York: Routledge, 2002), xi.

4. Alva Noë, *Action in Perception* (Boston: MIT Press, 2004), 1. Further, the way we grasp the world depends on the particulars of each sensual mode (vision, hearing, touch, smell) and the way of exploring the world that corresponds to each of these, giving rise to different patterns of "sensorimotor contingency," as Noë and J. Kevin O'Regan put it in "A Sensorimotor Account of Vision and Visual Consciousness," *Behavioral and Brain Sciences* 24, no. 5 (October 2001): 939–73. Thus, for example, a person born blind who was given vision late in life by the removal of cataracts expressed surprise that a circular coin turned about in the hand becomes thin and oblong when viewed on edge. He was also unfamiliar with the fact that distant objects appear smaller. Imagine how bizarre and disorienting this new visual experience must be for such a person. In exploring something with one's hands, how far one's hands are from one's trunk doesn't change one's perception of the size of the object being explored, and the "diskness" of a coin, for example, is invariant throughout the exploration. This invariant diskness is apparent by vision too (despite appearances, as it were) for a sighted person, but that is because such a person learned as a baby how a disk presents itself via reflected light from different angles, and this knowledge has become second nature.

    To suddenly establish a connection between the eyes and brain is merely to begin the long process of learning to see. And indeed, those who are given sight (in this bare sensual sense) after growing up blind report seeing nothing but blobs of light at first. Learning to see means mastering the pattern of sensorimotor contingencies by which reflected light maps, in a necessary and lawlike way, onto the physical objects that we encounter. These are "contingencies" because they hold only for animals like us who have two front-facing eyes. A duck has eyes that face to each side, so there is a different pattern by which *its* visual stimulation corresponds to our (shared) physical reality. Mastering these patterns, which we do as infants, depends on the ability to move around, just as the fingers must move around to perceive by touch. Only by exploratory action is one able to "extract invariants" from the scene, as James J. Gibson puts it.

5. James J. Gibson, *The Ecological Approach to Visual Perception* (Boston: Houghton Mifflin, 1979). One of the appealing things about Gibson's book is that it is fertile with suggestions about cognition that invite the reader to draw larger points, which Gibson himself refrains from making. When he writes that the "under-

lying invariant structure" emerges from the "changing perspective structure" (197), for example, one can't help but draw an analogy between visual perception and something like "insight." If self-motion is necessary to apprehend the world, this fits with our common intuition that wisdom is impossible without biography, that is, a self that moves or changes through time and thereby "gains perspective."

6. It follows that standard laboratory techniques of studying visual perception by putting someone in a headrest, having them look through an aperture, and then exposing them briefly to a two-dimensional picture in a darkened room, is to study only a very specialized case of vision, one that rarely occurs in everyday life and does not exploit the capacities we have developed as embodied creatures, both on the evolutionary time scale and developmentally. The proper technique would consist of providing optical information rather than imposing optical stimulation, as Gibson puts it.
7. Lawrence Shapiro, *Embodied Cognition* (New York: Routledge, 2011), 28.
8. As to the revolutionary status of Gibson: this is a convenient fiction, which I will go along with when it suits my immediate argumentative purposes. The current field of "ecological psychology" that takes Gibson as its inspiration may just as well be read as a working out of suggestions offered by Maurice Merleau-Ponty, Alfred Shutz, and Martin Heidegger—that is, by the school of phenomenology that formed about a century ago. The literature on embodied/embedded/grounded/extended cognition; ecological psychology; the move toward "enactivism"—in these literatures, one hardly ever sees references to the phenomenological writers, yet the most fruitful lines of argument emerging from them appear to be (largely surreptitious) borrowings from this philosophical tradition.
9. M. McBeath, D. Shaffer, and M. Kaiser, "How Baseball Outfielders Determine Where to Run to Catch Fly Balls," *Science* 268 (1995): 569–73, as cited by Andy Clark, *Supersizing the Mind: Embodiment, Action, and Cognitive Extension* (Oxford, UK: Oxford University Press, 2011), 16.
10. Clark, *Supersizing the Mind*, 35.
11. Ibid.
12. Gibson, *The Ecological Approach*, 128.
13. This way of life may be "predatory or preyed upon, terrestrial or aquatic, crawling or walking, flying or nonflying, and arboreal or ground-living" (ibid., 7). Unlike the proverbial brain in a jar, an embodied observer *has interests*, because it is capable of acting and being acted upon by objects in the world. Daniel Dennett makes the following argument for interestedness (and hence embodiment) as a precondition for the development of consciousness. The simplest animals recoil

from harm and approach what is good: food and opportunities to reproduce. These aversion and attraction behaviors enact the primordial categories of good and bad, and serve to guard against the dissolution of the boundary between self and nonself, i.e., death. The immune system enacts this principle at a mechanical level, where the distinction good/bad (for *me*) is encoded in an array of antibodies that react to molecules with particular shapes. If a creature is not able to move (some plants can turn, but cannot move their location), then the good/bad distinction hardly seems worth making. What would be the practical import of it? Such a creature simply has to take whatever impinges upon it. We can surmise further that such a creature would also have no use for consciousness, rooted in a distinction between self and nonself. According to this line of thought, the birth of subjectivity in animals is intimately bound up with a particular kind of embodiment: the kind that enables self-motion.

14. Noë, *Action in Perception*, 106.
15. I owe this example of the martial artist to William Hasselberger.
16. Adrian Cussins, "Experience, Thought and Activity," in Y. Gunther, ed., *Essays on Nonconceptual Content* (Cambridge, MA: MIT Press, 2002).
17. Here Clark (*Supersizing the Mind*, 45) is reporting the experiments of S. T. Boysen, G. Bernston, M. Hannon, and J. Cacioppo, "Quantity-based Inference and Symbolic Representation in Chimpanzees (*Pan troglodytes*)," *Journal of Experimental Psychology: Animal Behavior Processes* 22 (1996): 76–86.
18. Benedikt von Hebenstreit has studied vision and reactions in traffic. As Bernt Spiegel recounts it, von Hebenstreit conceptualizes three separate reaction circuits, corresponding to different parts of the brain: first, a reflexive circuit where "the processing of an external stimulus occurs in the spinal cord, on an extremely short pathway and thus very quickly, such as when one reflexively closes an eyelid or pulls away from sudden pain. These are inborn behaviors." The next "higher" level is the "subcortical reaction circuit," where those action programs that have been acquired but long since automated (such as riding a bicycle) run. The third and highest level is the cortical reaction circuit, in which "a stimulus has to not only be perceived but also interpreted to determine its meaning before an action is initiated in response—for example, when a child's ball rolls across the road in front of your car and you brake" (Bernt Spiegel, *The Upper Half of the Motorcycle: On the Unity of Rider and Machine* [Center Conway, NH: Whitehorse Press, 2010], 51). I believe the example of the ball rolling across the street is ideal here, as this is a relatively rare occurrence, and does indeed require an interpretive effort to respond appropriately. But I think the appearance of a red light, for example, does not fit this last category. You don't really perceive it as a mere sign, requiring in-

terpretation. Rather, the inhibition it stimulates is more immediate and visceral, because traffic lights are part of the normal routine of driving that we have been habituated to. If I am correct about this, it should be revealed by reaction time studies. We expect a green light to turn yellow, and then red, so our attention is never fully absent from traffic lights while driving, and the appropriate motor responses are primed—in a way that they are not for balls rolling across the road.

19. See Joseph K. Schear, ed., *Mind, Reason, and Being-in-the-World: The McDowell-Dreyfus Debate* (London: Routledge, 2013); and the back-and-forth between Dreyfus and McDowell in *Inquiry* 50, no. 4 (2007).
20. Spiegel, *The Upper Half of the Motorcycle*, 73. Spiegel's work was translated by Meredith Hassall from the original German *Die obere Hälfte des Motorrads: Uber die Einheit von Fahrer und Maschine* (Stuttgart: Heinrich Vogel Verlag, 1998).
21. An experienced motorcyclist is a bit like the gangster who is so cynical about human beings that he is never taken by a nasty surprise. His moral posture is adaptive, given his line of work—it lets him relax. He doesn't have to peer intently through the fog of other people's possible motivations and try to predict their behavior; instead he watches their hands. He is loose, preemptive, and ruthless.
22. Clark, *Supersizing the Mind*, 48.
23. This piano example is from D. Sudnow, *Ways of the Hand: A Rewritten Account*, as quoted by J. Sutton, "Batting, Habit and Memory: The Embodied Mind and the Nature of Skill," as quoted in turn by Clark, *Supersizing the Mind*, 237 n. 4.

3. VIRTUAL REALITY AS MORAL IDEAL

1. Thomas de Zengotita, *Mediated: How the Media Shapes Your World and the Way You Live in It* (New York: Bloomsbury, 2005).
2. In the opening scene of "Minnie's Mouseke-Calendar," the wind has blown away the pages of Minnie's day planner, which is quite distressing for her. The agenda that the young viewer is asked to identify with is that of the various institutions she is shuttled among in the course of her week; staying geared into the bureaucratic organization of time requires a coordinated calendar.
3. The first two quotes are from Immanuel Kant, *Groundwork for the Metaphysics of Morals*, ed. and trans. Allen W. Wood (New Haven: Yale University Press, 2002), 58–59, which corresponds to Ak 4: 440–41 in the canonical German edition. The last quote is from Immanuel Kant, *Groundwork of the Metaphysic of Morals*, trans. H. J. Paton (New York: HarperCollins, 2009), 108–109 = Ak 4: 441.
4. Indeed, the pleasure of demoting man from his special place seems to have supplied much of the psychic energy of early modern thought. Then as now, an en-

lightener is someone with the courage to live without illusion, face the truth of our condition, etc. It is a curiously self-aggrandizing form of humility.

5. Famously, Kant takes the golden rule—Do unto others as you would have them do unto you—and radicalizes it. One's own circumstances should not enter into one's deliberations; that would amount to special pleading. One must act in such a way that one could with good conscience distill the principle of one's action and turn it into universally binding legislation. It is central to Kant's moral doctrine that one should regard oneself and others as *representatives* of the generic category "rational beings," which may turn out to include Martians.

   Kant is after a *general theory* of morality, based on pure a priori reasoning—like arithmetic. That two plus two equals four is a fact that is impervious to experience; it will never have to be modified. In rejecting "accidental circumstances" and "the special constitution of human nature" as too parochial a basis for moral reasoning, Kant provides the clearest point of contrast to the idea of the situated self that animates this book.
6. Consider this passage in the *Groundwork*: "Even if it should happen that . . . this will should wholly lack power to accomplish its purpose, if with its greatest efforts it should yet achieve nothing, and there should remain only the good will (not, to be sure, a mere wish, but the summoning of all means in our power), then, like a jewel, it would still shine by its own light, as a thing which has its whole value in itself. Its usefulness or fruitlessness can neither add nor take away anything from this value" (*Groundwork*, trans. Paton, 62 = Ak 4: 394). Here Kant is saying the will can be pure, i.e., autonomous, *even if* impotent. But there would seem to be a closer affinity between purity and impotence than this "even if" formulation suggests. Purity is achieved by the same device of abstraction that guarantees impotence.

4. ATTENTION AND DESIGN

1. Eric Dumbaugh as quoted by Emily Anthes in "Street Smarts: Why White-Knuckle Roads Are a Driver's Safest Bet," *Psychology Today*, November/December 2010, 42.
2. This is the finding of Donald L. Fisher and Alexander Pollatsek in "Novice Driver Crashes: Failure to Divide Attention or Failure to Recognize Risks," in *Attention: From Theory to Practice*, ed. Arthur F. Kramer, Douglas A. Wiegmann, and Alex Kirlik (Oxford, UK: Oxford University Press, 2007), 149–50.
3. The automotive psychologist Bernt Spiegel did experiments in which he had sub-

jects walk along a ten-centimeter-wide black strip on the floor. The task was ridiculously easy, and seemed pointless to the subjects. But then it was revealed that the strip was actually a beam, which could be raised up off the floor. At a height of three or four centimeters, walking on the beam was still easy, but subjects concentrated on the placement of their feet. With the beam raised to the height of a chair, their gaze became fixed on the tiny area immediately in front of their feet, their arms spread out to the sides. At a height of four and a half meters, "there was now a physical tension that bordered on cramping." Bernt Spiegel, *The Upper Half of the Motorcycle: On the Unity of Rider and Machine* (Center Conway, NH: Whitehorse Press, 2010), 79.

4. Anthes, "Street Smarts," 42.
5. Arthur M. Glenberg, "What Memory Is For," *Behavioral and Brain Sciences* 20 (1997): 3.
6. In the field of signal processing, two signals are said to be time-locked if they share a common or absolute clock. One way to think about this is that the common clock guarantees that any hysteresis in one signal, indeed any departure from pure periodicity, is uniquely tied to the history of the other signal.
7. Andy Clark, *Supersizing the Mind: Embodiment, Action, and Cognitive Extension* (Oxford, UK: Oxford University Press, 2011), 17.
8. Paul Dourish, *Where the Action Is: The Foundations of Embodied Interaction* (Cambridge, MA: MIT Press, 2001), 102, as cited by Clark, *Supersizing the Mind*, 9–10.
9. A beautiful example of engineering that exploits our capacity for embodied cognition is the tactile flight suit. Fighter pilots face the problem of knowing which way is down, given the extreme g-forces their planes are capable of generating; they sometimes fly straight into the ground. The military developed a suit that indicates the direction of "down" with vibrations on the pilot's body. For helicopter pilots, one of the more challenging tasks is to keep the craft hovering in a stationary spot, especially if there are gusty winds. For them, the military has been developing a vest that emits a puff of air, which induces a tingling sensation, on the left side of the pilot's torso if the helicopter is rolling to the left, in the front if it is tilting forward, and so forth. Because the pilot is engaged in a goal-directed activity, these sensations become aversive, so he naturally moves his body in the opposite direction, away from the puffs. These natural reactions are used, in turn, to control the helicopter. Apparently the suit accomplishes a fairly effective integration of the helicopter with the pilot's body, such that he can fly blindfolded. He is no longer aware of the puffs of air, but rather of the pitch and roll of the helicopter. This is another case where a piece of equipment becomes transpar-

ent, and does so because there is a closed loop between action and perception, such that bodily motions affect sensory input, just as they do when we run to catch a fly ball.

Exclusive reliance on the visual display of information tends to produce information overload, and accordingly in environments where it really matters, there is growing interest in "multimodal interfaces" that make use of a variety of senses. In the intensely choreographed environment of a carrier flight deck, for example, visual and auditory attention are both severely taxed by the demands of communication and coordination, and so there too we find efforts to introduce the tactile presentation of information.

See Nadine Sarter, "Multiple-Resource Theory as a Basis for Multimodal Interface Design: Success Stories, Qualifications, and Research Needs," in *Attention: From Theory to Practice*, ed. Kramer, Wiegmann, and Kirlik, 188.

10. If you have ever listened to the NPR show *Car Talk* and heard people mimicking the sounds their cars make when they are misbehaving in some way, then you have some idea of the role played by sound in our ongoing monitoring of our cars, which we become aware of only when there is a new sound, indicating a problem.
11. Then again, it is said that we live at the end of history, so maybe we needn't fret about any of this. "In the future" (as Conan O'Brien used to say), we will be ferried around by Google's self-driving cars, wearing Google Glass goggles and who knows what all. The goggles will give us something exciting to watch, like *Grand Theft Auto*, and we will be given a steering wheel that shakes realistically as we execute brilliant evasive maneuvers. We will make *vroom vroom* sounds with our mouths to preserve that "sense of involvement," and arrive at our destination in a mood of triumph. We should have noted earlier that the passive kitten on the carousel has an enviable inner life.
12. Alfred Schutz and Thomas Luckmann, *The Structures of the Lifeworld*, trans. Richard M. Zaner and H. Tristam Engelhardt, Jr. (Evanston, IL: Northwestern University Press, 1973), 36–37.

5. AUTISM AS A DESIGN PRINCIPLE: GAMBLING

1. Natasha Dow Schüll, *Addiction by Design: Machine Gambling in Las Vegas* (Princeton, NJ: Princeton University Press, 2012), 171.
2. Mitzuko Ito, "Mobilizing Fun in the Production and Consumption of Children's Software," *Annals of the American Academy of Political and Social Science* 597 (2005), as cited by Schüll, *Addiction by Design*, 172.

3. Schüll, *Addiction by Design*, 174.
4. Ibid., 172.
5. Ibid.
6. Ibid., 12.
7. Caitlin Zaloom, "The Derivative World," *The Hedgehog Review* 12, no. 2 (Summer 2010): 20–27.
8. Schüll, *Addiction by Design*, 8.
9. Ibid., 56.
10. Ibid., 74.
11. Ibid., 87–88.
12. Ibid., 92.
13. Ibid., 224.
14. Ibid.
15. In the *Phaedo*, Plato's Socrates repeatedly says that philosophy is the practice of dying. I'm not sure how to take this. The fact that the hemlock is waiting just outside the door as he says it is probably relevant. But according to this strange motto, the gambling addict appears as the ultimate philosopher. He is moving beyond things that seem good—for him, as an individual—straight to the ultimate dissolution that is the end point of all organic life.
16. Schüll, *Addiction by Design*, 15.
17. And even in this narrow economic realm, we routinely fail to ask *Qui bono?*—Who benefits?—and refer instead to what is good or bad for "the economy" in the aggregate.

INTERLUDE: A BRIEF HISTORY OF FREEDOM

1. I owe this formulation of Locke's theological argument to Matt Feeney, personal communication.
2. Ibid. Perhaps an example—a digression on recent political history—can illustrate the abstract nature of the ideal of freedom established in the Lockean exercise, and the effects this ideal can have when unleashed on the world without regard for the particulars of an already established cultural setting.

   Political self-determination is an inspiring concept. So inspiring is it, and so resonant with our own self-image as Americans, that we eagerly recommend democratic forms to peoples with no history of such institutions, whose hearts and minds have not been habituated to self-government. Sometimes this recommending is done with the force of arms.

   If Edmund Burke were alive today, we could guess at how he would regard the occupation of and civil war unleashed in Iraq by a cell of democratic vanguardists in the Bush administration. Burke's pessimistic take on the French Revolution stemmed from his view that the French had imported the *doctrines* of English political liberty and set them down in an alien place. England had a centuries-long experience of the *practice* of liberty. It is a practice that requires skills, certain psychological dispositions, and a body of social knowledge that is mostly tacit. Political ideals take root in some soil that has been deposited over a long period, in some shared form of ethical life that provides for trust, and for the mutual intelligibility of citizens' actions and utterances. Absent such a context, Burke suggests, ideals are abstract and easily turn into their opposites. The Declaration of the Rights of Man becomes the Terror.

From our vantage a couple of centuries later, Burke's misgivings seem misplaced in the case of France. Iraq seems a better case for his argument, most importantly because democratic forms were imposed by an invading power.

Through their enthusiasm for other peoples' liberty that they expressed in the lead-up to the Iraq war, Republicans gave themselves over to a kind of self-regarding identity politics: we are not "appeasers." (The ghost of Neville Chamberlain was frequently invoked.) The main thing, for the vanguardist, is to savor the *clarity* of your own resoluteness in the service of freedom. The neocons may have wanted to be like Churchill, and to inspire those of military age to emulate the "greatest generation," but from our current vantage, the unseriousness with which they projected these psychic needs onto the world stage makes them look more like the radical-chic campus existentialists of the sixties.

Implicit in the project of evangelizing political self-determination is the whole anthropology of autonomy: a decisionistic exalting of the will, a corollary disregard of formation and character (we are free to create ourselves anew at each moment), and a privileging of maxims and ideals over practice. Recall the mood of Western journalists who triumphally disseminated photos of Iraqi women in burkas, each holding up an ink-stained finger to indicate that she had *voted.* That is, she had checked a box on a piece of paper, which pieces of paper were then collected by various mob bosses, for what purpose it is difficult to imagine, beyond the obvious one of stroking foreign observers by the mere act of collecting them. If I remember correctly, these photos adhered to a certain formula: a close-in shot of the smiling woman, somewhat abstracted from her surroundings by a shallow depth of focus. They were pictures of *individuals*, seemed to be the message, and this was very gratifying for us. It was a portrait of the liberal subject, fully realized in a flash through an act of political self-determination. We knew it was such an act because certain procedures were observed. Maxims and ideals are more easily formalized than social practices are, but the resulting forms often turn out to be empty and fragile. Such would be Burke's critique of Bush, that idealist.

3. Charles Taylor, *Sources of the Self: The Making of the Modern Identity* (Cambridge, MA: Harvard University Press, 1989), 169.
4. John Locke, *An Essay Concerning Human Understanding*, ed. Peter H. Nidditch (Oxford, UK: Oxford University Press, 1975), 101.
5. The term "self-responsibility" I have taken from Charles Taylor, who writes, "I have borrowed the term 'self-responsibility' from Husserl to describe something that Locke shares with Descartes (to whom Husserl applied the term) and which touches on the essential opposition to authority of modern disengaged reason" (*Sources of the Self*, 167).
6. Plato's Socrates had of course emphasized getting free of mere opinion and convention in order to arrive at the truth. But in principle one could be aided in this by some wise authority (in the parable of the cave there is a mysterious stranger who turns one around from the images projected on the wall by the poets, and leads one up to the sun). The point is to grasp an order that is independent of ourselves, and how you get to this point is not the important thing. The important thing is to turn one's attention from ephemeral, material things, and from mere images, to the unchanging Forms—from one set of external objects to another set of external objects. Once again, it is Charles Taylor who has clarified this contrast between ancient and modern thought on the question of where truth is to be found.
7. What he said, more precisely, was *"verum et factum convertuntur"*: the true and the made are convertible.
8. Martin Heidegger, "Modern Science, Metaphysics, and Mathematics," in *Basic Writings* (San Francisco: Harper and Row, 1977), 267–68.
9. There is an obvious strangeness here: from a beginning point that is radically self-enclosed (Descartes's "I think"), our task is to arrive at "a view from nowhere" (to use the philosopher Thomas Nagel's apt phrase) in which there remains no trace of the knower himself.

6. ON BEING LED OUT

1. Iris Murdoch, "The Sovereignty of Good over Other Concepts," in *Existentialists and Mystics: Writings on Philosophy and Literature* (New York: Penguin, 1999), 373.
2. Descartes, *Discourse on the Method for Rightly Conducting One's Understanding and for Seeking Truth in the Sciences*, trans. Donald A. Cress (Indianapolis: Hackett, 1980), 8 = p. 15 of the standard *Œuvres de Descartes*, ed. Adam and Tannery.

3. Immanuel Kant, "An Answer to the Question: 'What Is Enlightenment?'" in *Kant: Political Writings*, ed. Hans S. Reis, trans. H. B. Nisbet (Cambridge, UK: Cambridge University Press, 1991), 54.
4. In his later years, Emerson seems to have been a bit taken aback by the cultural effects of the very individualism he had been recommending. As against "the social existence which all shared," there "was now separation. Every one for himself; driven to find all his resources, hopes, rewards, society, and deity within himself." "There is an universal resistance to ties and ligaments once supposed essential to civil society. The new race is stiff, heady and rebellious; they are fanatics in freedom; they hate tolls, taxes, turnpikes, banks, hierarchies, governors, yea, almost laws." The last bit of this sketch sounds like today's Tea Party. Emerson as quoted by Wilfred M. McClay, *The Masterless: Self and Society in Modern America* (Chapel Hill: University of North Carolina Press, 1994), 55.
5. The quotes from Walt Whitman are from *Leaves of Grass* as quoted by McClay, *The Masterless*, 61. The quote from Norman Mailer is from "The White Negro," as quoted by McClay, 271.
6. McClay, *The Masterless*, 271–72.
7. Michael Polanyi, *Personal Knowledge: Towards a Post-Critical Philosophy* (Chicago: University of Chicago Press, 1974), 53.
8. Michael Polanyi, *The Tacit Dimension* (Chicago: University of Chicago Press, 1966), 20.
9. Polanyi, *Personal Knowledge*, 53.
10. Habit plays an important role. William James wrote, "As we become permanent drunkards by so many separate drinks, so we become . . . authorities and experts in the practical and scientific spheres, by so many acts and hours of work." According to James, the power of judging is built up silently "between all the details of [one's] business" (James, "Habit," *Principles of Psychology*, vol. 1, reprinted in *The Heart of William James*, ed. Robert Richardson [Cambridge, MA: Harvard University Press, 2012], 114). As the glassmaker knows when to fuse two molten pieces together, and cannot say exactly how, so the scientist has a sense of what line of inquiry is likely to be fruitful when he is faced with a novel problem.
11. Polanyi, *Personal Knowledge*, 53.
12. Ibid.
13. See the section entitled "What College Is For" in Matthew B. Crawford, *Shop Class as Soulcraft: An Inquiry into the Value of Work* (New York: Penguin, 2009), 143–48. I expressed reservations about some ways computers were affecting teaching in "The Computerized Academy," *The New Atlantis* (Summer 2005), available at www.thenewatlantis.com/publications/the-computerized-academy. And in "Sci-

ence Education and Liberal Education," *The New Atlantis* (Spring 2005), I argued that trying to motivate students to study science by emphasizing its social utility can only backfire. This last article is available at www.thenewatlantis.com/publications/science-education-and-liberal-education.

7. ENCOUNTERING THINGS WITH OTHER PEOPLE

1. Heidegger writes, "Strictly speaking there 'is' no such thing as *a* useful thing. There always belongs to the being of useful things a totality of useful things in which this useful thing can be what it is. A useful thing is essentially 'something in order to . . .' The structure of 'in order to' contains a *reference* of something to something else . . . [U]seful things always are *in terms of* their belonging to other useful things: writing materials, pen, ink, paper, desk blotter, table, lamp, furniture, windows, door, room. These 'things' never show themselves initially by themselves, in order then to fill out a room as a sum of real things . . . A totality of useful things is always already discovered *before* the individual useful thing." Martin Heidegger, *Being and Time*, trans. Joan Stambaugh (Albany: SUNY Press, 1996), 64.
2. Hubert Dreyfus gives the example of chopsticks as part of an equipmental whole in *Being-in-the-World: A Commentary on Heidegger's Being and Time, Division I* (Cambridge, MA: MIT Press, 1991).
3. Maurice Merleau-Ponty, *Phenomenology of Perception*, trans. Colin Smith (New York: Routledge, 2002), 36.
4. Alfred Schutz and Thomas Luckmann write, "The acquisition of knowledge is the sedimentation of current experiences in [already existing] meaning-structures, according to relevance and typicality. These in turn have a role in the determination of current situations and the explication of current experiences. That means, among other things, that no element of knowledge can be traced to any sort of 'primordial experience.' In analyzing the process of sedimentation which led to the development of the stock of knowledge, we always encounter prior experiences in which an already determined, albeit minimal, stock of knowledge must be conjoined." Alfred Schutz and Thomas Luckmann, *The Structures of the Lifeworld*, trans. Richard M. Zaner and H. Tristam Engelhardt, Jr. (Evanston, IL: Northwestern University Press, 1973), 119.
5. In a lecture course in Freiburg in 1923, Heidegger put it like this: "seeing arises out of and on the basis of a being-oriented regarding the objects, an already-being-familiar with these beings. Being-familiar with them is for the most part the sedimented result of having heard about them and having learned something

about them." Martin Heidegger, *Ontology: The Hermeneutics of Facticity*, trans. John van Buren (Bloomington: University of Indiana Press, 1999), 58–59.

6. The developmental account I offer here closely follows Jane Heal's in "Joint Attention and Understanding the Mind," in *Joint Attention: Communication and Other Minds*, ed. Naomi Eilan, Christopher Hoerl, Teresa McCormack, and Johannes Roessler (Oxford, UK: Oxford University Press, 2005), 38–39. The chronology is ambiguous as presented in this paper; in Eilan (cited in n. 8 below), gaze-following and gaze-checking are said to emerge only in the second year.
7. Christopher Mole, "Attention," in *The Stanford Encyclopedia of Philosophy*, ed. Edward N. Zalta (Fall 2009), available at http://plato.stanford.edu/archives/fall2009/entries/attention.
8. Naomi Eilan, "Joint Attention, Communication, and Mind," in *Joint Attention*, ed. Eilan et al.
9. Naomi Eilan writes that "Declarative pointings are not produced by chimpanzees, nor, usually, by autistic children. In both cases, moreover, researchers have strong intuitions that joint attention, in the sense of mutual awareness, is lacking." Ibid., 15–16. But see also D. Leavens and T. Racine, "Joint Attention in Apes and Humans: Are Humans Unique?," *Journal of Consciousness Studies* 16 (2009).
10. Heal, "Joint Attention," 39.
11. Axel Seemann, "Introduction," in *Joint Attention: New Developments in Psychology, Philosophy of Mind, and Social Neuroscience* (Cambridge, MA: MIT Press, 2011), 2.
12. This is the position of the growing school of "interactionism" that straddles social psychology and cognitive psychology. In coordinated action, there is "a 'meeting' of minds rather than an endless ascription of higher-order mental states," as Mattia Gallotti and Chris D. Frith put it in "Social Cognition in the We-Mode," *Trends in Cognitive Sciences* 17, no. 4 (April 2013): 164.
13. Eilan, "Joint Attention," 2.
14. See Gary L. Wells, "Eyewitness Identification: Systemic Reforms," *Wisconsin Law Review*, 2006, 615–43.
15. There are subtleties here. Descartes is talking about beliefs, which is one class of what contemporary analytical philosophy calls propositional attitudes. The point I have been making in this chapter (and in "Embodied Perception") is mostly about nonpropositional mental content: our grasp of sensorimotor contingencies; the phenomenological priority of social facts over sensual ones; the salience of equipmental norms as against bare physical affordances. But I believe these points about our prearticulate mental life go a long way toward showing why the ideal of epistemic individualism, which is offered in the domain of

articulate propositional attitudes such as beliefs, is an impossible one. One can't build such attitudes for oneself, from the ground up (as Descartes tries to do) because there is so much social determination of our mental contents prior to articulate propositions.

8. ACHIEVING INDIVIDUALITY

1. Robert B. Pippin, *Hegel's Practical Philosophy: Rational Agency as Ethical Life* (Cambridge, UK: Cambridge University Press, 2008), 173.
2. Kant had tried to square freedom with adherence to universal norms by conceiving autonomy to lie in the moment when you choose to act according to those norms, which you have apprehended through reason. This emphasis on the rational will's moment of choice probably laid the ground for French existentialism, which jettisoned the bit about rational norms, leaving the subject to make a groundless choice. Indeed, to see and accept the groundlessness of the choice is the mark of courage or probity. A person's value judgments are not tied onto the world, they do not apprehend anything, so the idea of "doing the right thing" loses its force. In fact, that idea amounts to a failure to grasp one's radical responsibility for oneself. Value judgments are pure flashings forth of the personal will, and the main thing is to hold them sincerely.

   The valor of the choosing will is thought to derive from the fact that its choice is so *significant*. Should I go off to fight with the resistance or stay and take care of my dying mother? (This is Sartre's famous example.) Both "values" are said to rest on a radical choice. But, as Charles Taylor argues, this situation is felt as a dilemma only because the competing claims on the young man are *not* created by radical choice. "If they were[,] the grievous nature of the predicament would dissolve, for that would mean that the young man could do away with the dilemma at any moment by simply declaring one of the rival claims as dead and inoperative. Indeed, if serious moral claims were created by radical choice, the young man could have a grievous dilemma about whether to go and get an ice cream cone, and then again he could decide not to." Charles Taylor, "What Is Human Agency?" in *Human Agency and Language: Philosophical Papers* 1 (Cambridge, UK: Cambridge University Press, 1985).
3. But note that even in a "free market," my customer and I do not come to the exchange on a footing of equality; I have superior information. This asymmetry tempts me to be dishonest. Only to the extent I resist this temptation can I take the customer's payment of the full bill as validation of the claim I have made for

myself. Even if I am completely honest, however, this satisfaction is of a lower order than, or at any rate different from, that which I would receive from another mechanic who recognizes the work as being well done.

4. Matt Feeney, personal communication.

9. THE CULTURE OF PERFORMANCE

1. Alain Ehrenberg, *The Weariness of the Self: Diagnosing the History of Depression in the Contemporary Age* (Montreal: McGill-Queen's University Press, 2010), 4.
2. Ibid., 218–19.
3. See Richard Sennett, *The Corrosion of Character* and *The Culture of the New Capitalism.*
4. Pew Charitable Trusts, "Economic Mobility Project Fact Sheet: Does America Promote Mobility As Well As Other Nations?," available at www.pewstates.org/research/reports/does-america-promote-mobility-as-well-as-other-nations-85899380321. See also Joseph E. Stiglitz, "Equal Opportunity, Our National Myth," *The New York Times,* February 16, 2013, available at http://opinionator.blogs.nytimes.com/2013/02/16/equal-opportunity-our-national-myth. The quote about Americans' faith in the meritocracy is from Howard Steven Friedman, "The American Myth of Social Mobility," *The Huffington Post,* July 16, 2012, available at www.huffingtonpost.com/howard-steven-friedman/class-mobility_b_1676931.html.
5. Alexis de Tocqueville, *Democracy in America,* trans. Henry Reeve (New York: Colonial Press, 1899), 2: 104–106, as quoted by Wilfred M. McClay, *The Masterless: Self and Society in Modern America* (Chapel Hill: University of North Carolina Press, 1994), 43–44.
6. After the financial crisis of 2008 that inaugurated the Great Recession, the Tea Party arose. One way to read this is that we quickly inoculated ourselves against the conclusions that threatened to be drawn from the crisis—the revelation of fundamental *antagonisms* in the economy—by doubling down on the very individualism that had obscured such developments from our sight. Consider the right's response to President Obama's "You didn't build that" speech. Obama was addressing a certain myopic egotism that we all fall into sometimes. He pointed out that if you are successful, you probably received some help along the way—from a teacher who took an interest in you, for example. Any business that thrives depends on a background of public investments (roads, police) that make it all possible, but which are easy to take for granted. I took the president to be saying, "You didn't build *that*!" It seemed a fairly commonsensical reminder that there is

something called the common good. But this gets called socialism. It is the "risk-takers" who generate all prosperity, ex nihilo. Never mind that Romney built his own fortune risking *other* people's money and livelihoods. And this brings up an important point. However legitimate the role of venture capital, leveraged buy-outs, and financial services in society, these activities are very different from small business, and the Republican Party of 2012 seemed to take as one of its most important tasks the clouding of this difference, so as to be able to attach the moral valor of the entrepreneur to enterprises that primarily capture wealth. The right's schema of "makers versus takers" is perhaps apt enough, if taken in a way that is politically opposite to how it is used on Fox News. The utility of this phrase on the right in the 2012 election depended on reviving the 1980s figure of the welfare queen, who had more or less disappeared from our political imagination—and political reality—with welfare reform under Clinton. She and her paltry food stamps are back, because she is indispensable to the rhetorical job Republicans have set themselves, namely, distracting attention from the more massive, structural (and indeed state-enabled) transfers of wealth that are passed under cover of "free market principles."

7. This way of redescribing our experience, from a brain perspective that stands outside experience, is closely allied to the popular image of artificial intelligence. If the human being is fully describable in material terms, the thought goes, we should be able to replicate a person's consciousness in any physical system that is sufficiently rich and stable to capture all the neurological connections that constitute it. The irony of this kind of materialism is that it treats the human being *not* as material, but as pure form: an information system, the decisive features of which float free of any particular material instantiation. Either a synapse fires or it doesn't, and like any binary machine state this can be represented as a zero or a one. What is important is the logical relations between parts of the system. The argument has been made that artificial intelligence is a Kantian project: it seeks to establish the a priori conditions for the possibility of knowledge, in strictly universal terms that would apply to "all rational beings" whatsoever, whether they happen to be human, electronic, or an abacus made out of empty beer cans. Reductive materialism in the human sciences usually takes itself to be the hardheaded opposite of idealism, but is better understood as a species of idealism. The material objects it is concerned with are abstractions, conceived as logical units isolated from the contingencies that affect any actual, embodied being, or indeed any actual machine. Warren McCulloch was one of the group of thinkers who launched the research program of cybernetics. According to Jean-Pierre Dupuy, McCulloch referred insistently to Kant in his writings, and sought to establish "a physical

basis for synthetic a priori judgments." The Kantian character of AI was asserted by Joël Proust in "L'intelligence artificielle comme philosophie," *Le Debat*, no. 47 (Nov.–Dec. 1987). I have relied on Dupuy's account of Proust's argument in his outstanding book *The Mechanization of the Mind: On the Origins of Cognitive Science*, trans. M. B. DeBevoise (Princeton, NJ: Princeton University Press, 2000), 93.

10. THE EROTICS OF ATTENTION

1. Transcription of the 2005 Kenyon College Commencement Address written and delivered by David Foster Wallace, May 21, 2005, available at http://web.ics.purdue.edu/~drkelly/DFWKenyonAddress2005.pdf.
2. To be sure, generous imaginings are not the same as beliefs; one can actively entertain them without putting money on them, as it were, and this is probably a good thing to try to do as a corrective to our tendency to regard others as obstacles to our will, especially in the kind of mute encounters Wallace describes, where one has no idea what others' backstory might be.
3. Iris Murdoch, "On God and Good," in *Existentialists and Mystics: Writings on Philosophy and Literature* (New York: Penguin, 1999), 345.
4. Ibid., 344–45.
5. Thus Adam Seligman and his coauthors have argued in *Ritual and Its Consequences: An Essay on the Limits of Sincerity* (Oxford, UK: Oxford University Press, 2008). The example of spousal ritual I give here is taken from a lecture delivered by Seligman at the University of Virginia's Institute for Advanced Studies in Culture on November 18, 2009.
6. William James, "The Gospel of Relaxation," in *The Heart of William James*, ed. Robert Richardson (Cambridge, MA: Harvard University Press, 2012), 132, emphasis added. In this popular essay James is giving practical advice that is based on the James-Lange theory of emotions, according to which feeling tends to follow action.
7. I owe this insight to a conversation with Talbot Brewer.
8. Sherry Turkle, *Alone Together: Why We Expect More from Technology and Less from Each Other* (New York: Basic Books, 2011), 177.
9. Ibid., 10.

11. THE FLATTENING

1. David Brooks, "If It Feels Right," *The New York Times*, September 12, 2011.
2. This point is made by Talbot Brewer in *The Retrieval of Ethics* (New York: Oxford University Press, 2011).

3. I owe the argument of this paragraph, and some of the language, to conversations and email exchanges with William Hasselberger and Talbot Brewer. They, in turn, report that they have been informed by Cora Diamond, " 'We Are Perpetually Moralists': Iris Murdoch, Fact, and Value," in *Iris Murdoch and the Search for Human Goodness*, ed. Maria Antonaccio and William Schweiker (Chicago: University of Chicago Press, 1996).
4. I owe this point to a conversation with Daniel Doneson.
5. Kant's insistence on universality of aesthetic judgment demands that we "avoid the illusion arising from subjective and personal conditions which could easily be taken for objective, an illusion that would exert a prejudicial influence on [one's] judgment." Immanuel Kant, *Critique of Judgment*, trans. James Creed Meredith (Oxford, UK: Oxford University Press, 1952), sect. 40, 151–52, as quoted by Adam Adatto Sandel in his excellent book *The Place of Prejudice: A Case for Reasoning Within the World* (Cambridge, MA: Harvard University Press, 2014), 49. These "subjective and personal conditions" arise from your situation—from your own biography, as someone who has been formed in some historically contingent local context that you didn't really choose. Maybe you grew up listening to Southern California hard-core punk in the 1980s because that's what your friends were into. Anything parochial like that is a source of prejudice to be rooted out, according to Kant.
6. Kant, *Critique of Judgment*, sect. 40, 151.
7. Sophia Rosenfeld writes that "in Kant's telling, in judging aesthetic matters we become unusually aware of our links to others." Aesthetic judgment holds forth the potential for "agreement founded on affective identification with the other." *Common Sense: A Political History* (Cambridge, MA: Harvard University Press, 2011). I find this version of Kant attractive, but hard to square with Kant's insistence that we refer ourselves to a generic "everyone" for a check on our own aesthetic response; that strikes me as at odds with any actual experience of aesthetic communion with others.
8. I owe this point about Arendt to Matt Feeney (personal communication).
9. Søren Kierkegaard, *The Present Age: On the Death of Rebellion*, trans. Alexander Dru (New York: HarperCollins, 2010), 33–34. This essay was originally published in 1846.
10. Ibid., 25.
11. Ibid., 15–16.
12. Wilfred McClay, "The Family that *Shoulds* Together," *The Hedgehog Review*, Fall 2013, 25–26.
13. Let us pause to consider how the inequalities that are rife in American society

relate to the kind of differentiation that Kierkegaard defends. One problem with our big disparities of wealth is that (as Aristotle said in the *Politics*) plutocrats always think that superiority in this one dimension (wealth) is an indication of superiority in every dimension, and therefore a title to rule. In our time, we call this assumption "meritocracy." (Mitt Romney was certain he was "the smartest guy in the room" because he was the richest guy in the room.) Because this pretense is so patently false, I think it makes us suspicious of all kinds of hierarchy as having some kind of bad faith at its origin. (But note our ready recognition of athletic excellence; we are not cynical about sports.) This is a stance that flatters our egalitarian self-satisfaction, and probably contributes to the flattening that Kierkegaard laments.

14. Kierkegaard, *The Present Age*, 16–17, emphasis added.

12. THE STATISTICAL SELF

1. Sarah E. Igo, *The Averaged American: Surveys, Citizens, and the Making of a Mass Public* (Cambridge, MA: Harvard University Press, 2007).
2. H. R. Markus and S. Kitayama, "Culture and the Self: Implications for Cognition, Emotion, and Motivation," *Psychological Review* 98 (1991): 224–53. The Japanese and other Asian peoples are more likely to give concrete, role-specific descriptions of themselves in particular settings, such as being at home or at work, while Americans describe themselves with "more psychological trait or attribute characterizations ('I am optimistic,' and 'I am friendly')" (ibid., 233).
3. Igo, *The Averaged American*, 285.
4. Ibid., 284.
5. Kinsey also shined a bright light on the prevalence of marital infidelity. He probably hoped this would encourage a more humane and lenient standard of marriage, but I suspect it had the effect of helping to destroy any space for privacy within marriage, as though spouses should not have anything to hide from each other. It seems plausible to trace a connection between the social scientific revelation that your spouse, if he or she is representative, is likely fooling around on you, and the fact that the divorce rate shot up in the third quarter of the century. Monogamy had been a thin yet socially necessary pretense; it was exploded in the name of sincerity and replaced with a deeper, more demanding sexual moralism of "honesty" that has turned out to be unlivable for many people.
6. This is the same university where, during one stretch, nearly every other student at the library seemed to be reading a book entitled *How to Think Like a CEO*. One

shouldn't allow oneself to be dismayed by such things. The university's mission of "preparing students for life" may actually require uneducating them, so they will be better adjusted. If they read this book instead of Shakespeare, they won't feel oppressed by the impoverished language they are expected to use at work. And if they can identify with the CEO, they will be less likely to feel themselves in an antagonistic relation to those who manage the appropriation of their surplus labor value on behalf of Chinese shareholders.

7. Jaron Lanier, "Digital Maoism: The Hazards of the New Online Collectivism," *Edge*, May 29, 2006, available at http://edge.org/conversation/digital-maoism-the-hazards-of-the-new-online-collectivism.

13. THE ORGAN MAKERS' SHOP

1. I imagine the appeal of this image may have something to do with the fact that it allows the venture capitalists who hang around Silicon Valley to view themselves in a certain cultural role, as midwives to the new. This is like being a patron of the avant-garde: quite apart from any profit that may come, one has the sense of being in touch with the most important experiments under way, the most radical possibilities. This is an exceptional position to be in at a cocktail party.
2. "When viewing the Riverside Church console for the first time, Monsieur put his hand to his head and said, 'Mon Dieu!,' while all Madame could say was 'Oo la la!' " From the recollections of John Tyrrell, former president of the Aeolian-Skinner Organ Company, http://aeolian-skinner.110mb.com.
3. My site visits to Taylor and Boody were in 2007 and 2008. The conversations I report here are based on recordings I made during those visits.
4. According to Florensky, post-Renaissance Catholicism entailed "the distortion of the whole of spiritual life . . . And the essential sonic expression of this Catholicism is the organ sound." Noting that Renaissance oil painting developed and flourished alongside the art of building pipe organs and composing music for them, he finds in these "two material causes arising from the same metaphysical root." In particular, "The very consistency of oil paint has an obvious affinity with the oily-syrupy sound of the organ; and the flatness and liquidity of oil colors inwardly connects them to the sonic liquidities of the organ. Both the colors and the sounds are wholly of earth and flesh." Pavel Florensky, *Iconostasis*, trans. Donald Sheehan (Oakwood Publishing, 1996), 100–102. Consider a more recent parallel: the early critics of rock and roll saw in it a threat to civilization. And of course, they were right. In a sense they held the music in higher regard—took it more

seriously—than the appreciative rock critics of later decades who grew up with the music and lived within its horizon. In the same way, this hostile reaction to the organ as a work of the flesh perhaps serves as a clarifying counterweight to any too-easy appreciation of the organ as something innocuously "spiritual."

5. As quoted in Lawrence Phelps, "A Short History of the Organ Revival," *Church Music* 67 (1967).
6. In a letter to Joseph S. Whiteford, president of the Aeolian-Skinner Organ Company, dated March 27, 1960, Albert Schweitzer writes, "The sound of today's organs is in grave danger, because the architects do not provide the location for them where their tonal effect would be best, and because much sound-absorbing material is being used in churches nowadays, which swallows the sound and destroys the magnificent resonance which stone gives to the sound of the organ." Charles Callahan, ed., *Aeolian-Skinner Remembered: A History in Letters* (n.p.: Randall M. Egan, 1996).
7. Lawrence I. Phelps, "Trends in North American Organ Building," delivered at the International Organ Festival, St. Albans, June 28, 1969, and published in *Music: The AGO & RCCO Magazine*, May 1970. See also Phelps's article "A Short History of the Organ Revival," *Church Music* 67 (1967).
8. We can now see that there may have been a historical error in the complaint of the Orthodox theologian about the sound of the organ (note 4 above). Florensky associates the hated sound with the Baroque period of organ music, but the actual sonic qualities he was reacting to may well have been those of a later epoch, produced by the Romantic orchestral organ that was prevalent in Florensky's time, before the Baroque revival (it would be interesting to know what particular organs he was reacting to). His description of the organ as "slow, submerged," "flat and liquid," and "oily-syrupy" is practically identical to how organ builders themselves describe the sound of the Romantic orchestral organ, and it was in reaction against this sound that they set out to recover the principles of Baroque organ making, resulting in a sound that is much harder, crisper, and clearer. The irony is that the recovery of clean voicing in organs was made possible by the examination of organs dating from the period Florensky identifies as the problem, in his determination to identify Renaissance humanism as the source of all decadence.
9. In a letter to Henry Willis, dated February 16, 1949, the organ builder Ernest Skinner writes, "The way pneumatic leather (sumac) is tanned now, it is only good for about 18 years." Charles Callahan, ed., *The American Classic Organ: A History in Letters* (Organ Historical Society, 1990), 281.
10. George A. Audsley, *The Art of Organ-Building*, vol. 2 (New York: Dover, 1988), 623–24.

11. This polemical opposition was set up quite self-consciously by early modern thinkers such as Bacon, Hobbes, and Machiavelli. The "relief of man's estate" through scientific progress required freeing men's minds from the darkness of "the schools."

EPILOGUE: RECLAIMING THE REAL

1. Doug Stowe, "Wisdom of the Hands" (blog), October 16, 2006, available at http://wisdomofhands.blogspot.com.

ACKNOWLEDGMENTS

Once a week I step into clammy Aerostich gear and ride seventy-five miles on beautiful country roads to Charlottesville. There I stop by the office of my friend Joe Davis, publisher of *The Hedgehog Review*, ideally when he is not there. Whatever book is on his desk, I take. The book you have just read is simply a record of my efforts to synthesize these takings over the last few years. It follows that whatever shortcomings this book may have are due to defects in Joe's taste.

I have some really smart friends. But lots of people are smart. More important, I have been fortunate to know people who share a certain sensibility, a dispositional discomfort with the times that gives their intellectual inquiries a more personal inflection than is usual in the scholar.

Matt Feeney, Tal Brewer, and Bill Hasselberger generously agreed to hole up with me at a cabin in the Blue Ridge Mountains in 2013 to work through what I thought was a complete draft of the book (and my best liquor). The result was another seven months of work on it. One upshot of these conversations is that Kant started to emerge as the thinker I had been shadowboxing with, without being fully aware of the fact. The role that he subsequently came to play in the book—the influence over our everyday psychology that I assign to him—became a matter of some contention among us. Tal tried valiantly to disabuse me of my reading of Kant, without much success. This entailed some obstinacy on my part, as Tal was once a serious student of Kant, whereas I am not now nor have I ever been. But in reading the *Groundwork*, belatedly, I felt I was reading a crystallization, or rather an over-the-top parody, of the psychology of freedom I had been criticizing. That book also revealed the deep connection between our stance toward the world of things and the kind of moralism that flies the flag of the self. Kant thus made more explicit for me the conceptual link between Parts I and II of my own book. The flood of gratitude that this inspired in me toward the august East Prussian philosopher was such that I had no choice but to seat him at the head of the table and allow him to speak. Tal thinks I put a whoopee cushion under him; I think he needs no such device.

Feeney, too, is a former Kant scholar, more sympathetic to the link I try to establish between Kant and creepy children's television. He helped me place the *Groundwork* in the context of Kant's vast system. Few of the subtleties and qualifications Feeney impressed on me survived the editing process; what survived was the energy of our exchange. More generally, Feeney has for years been that one reader the existence of whom makes it possible for me to write. One has to know that there is someone whose gut is populated by the same flora, setting off similar immune reactions to the cultural biome. Our overlapping educations in the history of political thought have similarly equipped us to understand these reactions as having reasons to them, which are recoverable with some genealogical effort.

Bill helped me to see the theory of human action that is implicit in Kant's ideal of autonomy, and to draw an explicit contrast between it and the view of action that is contained in the notion of affordances. This was key.

Daniel Doneson brought a mood of Socratic urbanity to Charlottesville, which that town sorely misses in his absence. If they were wise, the city fathers would establish an endowment for his maintenance (in the lavish style he requires), for he single-handedly made the downtown pedestrian mall a place for intellectual intrigue and the pure pleasure of conversation.

Beth Crawford was an indispensable guide to the research on

embodied cognition and other currents in cognitive science that I would have been unaware of otherwise, and provided a critical sounding board for my working out of many of the ideas here. My debt to her is a large but happy one.

My erstwhile business partner John Ryland of Classified Moto was understanding when the protracted writing of this book kept me from the joint venture we had planned. Anyone looking for a custom motorcycle should look him up.

I thank Natasha Dow Schüll and Princeton University Press for permission to quote extensively from *Addiction by Design*.

Mike Rose of UCLA, an indispensable gadfly to our education system, gave me valuable comments on the chapter "On Being Led Out," as did Peter Houk of MIT. (Peter also gave me two beautiful specimens of his glassblowing—fossilized bits of joint attention ideally suited to drinking Scotch out of—that I cherish.) Ty Landrum gave a response to "Attention as a Cultural Problem" when I presented it at the University of Virginia's Institute for Advanced Studies in Culture in 2010. Brie Gertler's intervention brought the problem of self-knowledge into focus for me at a crucial stage, and her sincere compliments had a hugely renewing effect. The craftspeople at Taylor and Boody were very generous in tolerating my presence and answering my questions in 2007 and 2008. In particular I would like to thank John Boody, Chris Bono, Chris Peterson, Ryan Albashian, Kelly Blanton, Tom Karaffa, and Robbie Lawson, as well as the organist (and family friend) Frank Archer.

It is hard for me (and surely for the publisher) to anticipate how this book will be received. The thread of the inquiry required imposing arguments of some intricacy on readers who are often assumed to have little patience for the effort they require. Eric Chinski, my editor at FSG, provided me a luxurious space in which to let the book take the course that it did, based on his thorough comprehension of what I was trying to accomplish. I count myself extraordinarily lucky to have found this friend and benefactor in the publishing world.

Will Hammond, my editor in the UK, brought a rare level of seriousness to the project and provided countless suggestions that improved the work. My agent, Tina Bennett, has been the best sort of colleague, equal parts intellectual companion and fiduciary tigress. The enthusiasm of professionals as talented as Will and Tina provided me a reassuring point of triangulation when the book seemed too ambitious and unusual to have a chance of being read out there, in the world beyond my own head.

Upon Will's departure from Viking UK, Daniel Crewe inherited the book and loved it as his own, and for this I am grateful.

Finally, my fellowship at the Institute for Advanced Studies in Culture has been a boon. Due to what I assume must be a clerical error, they have continued their support while asking nothing of me. Big thanks to James Hunter, Joe Davis, and the whole crew for making me part of their conversation, which has enriched my own thinking immeasurably.

INDEX